# SELF-STUDY WORK

# CHEMISTRY

## AN INTEGRATED APPROACH

### CATHERINE E. HOUSECROFT
### EDWIN C. CONSTABLE

LONGMAN

**Addison Wesley Longman Limited**
Edinburgh Gate
Harlow
Essex CM20 2JE
England
and Associated Companies throughout
the World.

ISBN 0 582 27430 3

**British Library Cataloguing-in-Publication Data**
A catalogue record for this book is
available from the British Library

Printed in Great Britain by Henry Ling Ltd.,
at the Dorset Press, Dorchester, Dorset

**Also available**
Chemistry: An Integrated Approach          ISBN 0582 25342 X

# Contents

# Preface

This *Self-Study Workbook* accompanies *Chemistry: An Integrated Approach*. The book is not simply a series of questions. It has been designed to stimulate students into working on their own and we have provided more detailed answers than space allowed us in the main text. Indeed, one-quarter of the workbook is given over to answers. The workbook is intended for use both by individual students and small study-groups.

The chapter organization of the *Self-Study Workbook* parallels that of *Chemistry: An Integrated Approach* and enables students to find appropriate reading material and problem sets to support lecture material. Its design also makes the book suitable for use either within a first year general chemistry course or to accompany separate first year courses in organic, physical and inorganic chemistries. In most chapters we have introduced additional topics which represent review material from pre-university (e.g. Graham's Law and the kinetic theory of gases in Chapter 1), extensions of discussions from the main text (e.g. chirality and optical activity in Chapter 8, indicators and buffers in Chapter 11, and the theory of orientation effects, activation and deactivation of aromatic rings in Chapter 15), or applications which combine material from two or more chapters (e.g. emissions of NO and $SO_2$ in Chapter 13, and multi-step organic syntheses in Chapters 14 and 17).

An important aspect of the workbook is the extensive cross referencing to *Chemistry: An Integrated Approach*. This has two aims. Firstly, rather than give all of the data necessary for a problem, we ask students to find appropriate physical data from the Appendices in *Chemistry: An Integrated Approach*; this exercise is designed to encourage students to think more carefully about the data needed. Secondly, we cross-reference discussions in the main text that are relevant to the problem-solving exercises.

A workbook of this kind must be tried and tested before publication. We are grateful to the members of our research group who have tirelessly worked through the chapters, checking the answers and providing criticism and suggestions for revisions. In particular we thank Sarah Angus-Dunne, Simon Dunne, Torsten Kulke, Emma Schofield and Diane Smith.

The *Self-Study Workbook* has been prepared and typeset by the authors using Adobe Pagemaker v. 6.0; the majority of the graphics were constructed using ChemDraw Pro, Chem3D Pro or Cricket Graph III v. 1.5. We must extend our thanks to those at Addison Wesley Longman for their assistance and seemingly never ending e-mail messages and faxes reminding us that the deadline was nigh and the page limit could not be extended any further; special thanks go to Jane Glendening and Alex Seabrook.

Authorship within our house is never without the assistance of Philby and Isis; most of the text has been written around two Siamese cats sleeping or playing close to the warmth of the Macintosh. We have much to thank them for in terms of keeping our lives in perspective.

Although many people have read the text carefully, some errors (although we hope not many) will inevitably remain. If you have suggestions for future editions of this workbook, please let us know:

HOUSECROFT@UBACLU.UNIBAS.CH
CONSTABLE@UBACLU.UNIBAS.CH

Catherine E. Housecroft
Edwin C. Constable

Basel
December 1996

## Important note to readers

Through this workbook, the abbreviation **H&C** stands for **Housecroft & Constable,** *Chemistry: An Integrated Approach*.

# 1 Some basic concepts

---

**Topics**

- Atomic number, mass number, isotopes and mass spectrometry
- The mole, Avogadro constant and relative molecular mass
- Gas laws, ideal gases and the kinetic theory of gases
- Solution concentration
- Stoichiometry and balanced equations
- Oxidation states and balanced redox reactions
- Enthalpy changes and Hess's Law
- Equilibria

---

## ATOMIC NUMBER, MASS NUMBER, ISOTOPES AND MASS SPECTROMETRY

The atomic number of an atom gives the number of protons and the number of electrons in that atom:

Atomic number, $Z$ = Number of protons = Number of electrons.

The difference between the mass number and the atomic number gives the number of neutrons:

Mass number = (Number of protons) + (Number of neutrons).

Isotopes of an element are atoms in which there are different numbers of neutrons; the number of protons and the number of electrons are constant for a given element. Isotopes can be observed by using *mass spectrometry* (Figure 1.1). The sample to be analysed is first vaporized by heating and is then ionized. For an element which is a solid at 298 K, these processes are shown in equations 1.1 and 1.2 although it is also possible for multiply charged ions to be formed.

**1.1** A schematic representation of a mass spectrometer. The sample shown here produces two ions after passing through the ionizing chamber and these are separated by the magnetic field.

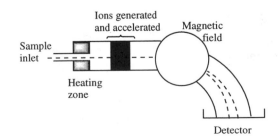

$$E(s) \xrightarrow{\Delta} E(g) \tag{1.1}$$
$$E(g) + e^- \rightarrow E^+(g) + 2e^- \tag{1.2}$$
$$\underset{\text{fast}}{\phantom{E(g)}} \qquad \underset{\text{slow}}{\phantom{E^+(g)}}$$

The ions pass through a magnetic field where their path is deflected; the amount of deflection is mass dependent. The detector output is called a *mass spectrum* — a plot of signal intensity against magnetic field strength which can be calibrated to give a plot of signal intensity against mass:charge ($^m/_z$) ratio. For a *singly* charged ion, this ratio is equal to the mass. Figure 1.2a shows the mass spectrum of naturally occurring *atomic* carbon which consists of 98.9% $^{12}C$ and 1.1% $^{13}C$. The effects of having $^{13}C$ present in a sample of carbon *atoms* are not great, but they become significant in a molecule which contains many carbon atoms. Figure 1.2b shows

**$C_{60}$: see H&C Section 7.9**

part of the mass spectrum of the fullerene $C_{60}$. The peak at $^m/_z$ 720 corresponds to $C_{60}$ containing all $^{12}C$ atoms, whilst that at $^m/_z$ 721 is assigned to $(^{12}C)_{59}(^{13}C)$. The chance of finding one $^{13}C$ atom in a molecule of $C_{60}$ is high. [*Exercise:* How can you rationalize the $^m/_z$ values and peak intensities for the remaining peaks? Why is no peak visible higher than $^m/_z$ 724?]

**1.2** (a) A representation of the mass spectrum of atomic carbon; the *relative* intensity scale sets $^{12}C$ at 100. (b) The highest mass peaks in the mass spectrum of $C_{60}$.

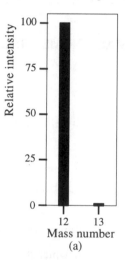

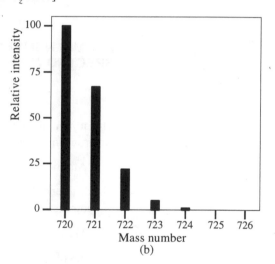

Figure 1.2b is only *part* of the mass spectrum of $C_{60}$. Many molecules fragment within the mass spectrometer and signals due to these fragments are also observed. The highest mass peaks usually correspond to ions formed from the complete molecule but with different combinations of isotopes.

**Problem set 1.1**

**Data needed are listed in H&C Appendix 5**

1. Sketch the isotope distribution of the group of highest mass peaks in the mass spectrum of (a) ClNO; (b) $Br_2$; (c) $XeF_2$.
2. What peaks would you expect to see in the mass spectrum of $F_2$?
3. The relative molecular masses of $CH_3CH_2SH$ and $HOCH_2CH_2OH$ are both 62. How could mass spectrometry be used to distinguish between them?
4. Determine the relative atomic mass of potassium to 4 significant figures.
5. In the mass spectrum of $CCl_4$, the highest mass peaks are at $^m/_z$ 152 (78), 153 (1), 154 (100), 155 (1), 156 (48), 157 (1), 158 (10), 160 (1) where the

numbers in parentheses are the relative intensities of the peaks. Assign each peak. What other peaks might be expected to be present in the mass spectrum of $CCl_4$?

## THE MOLE, THE AVOGADRO CONSTANT AND RELATIVE MOLECULAR MASS

$L = 6.022 \times 10^{23} \text{ mol}^{-1}$

This section reviews a range of calculations dealing with mass, moles, relative molecular mass and the Avogadro constant, $L$.

**Worked example 1.1**

**Determine the mass of 0.25 moles of $K_2SO_4$ ($A_r$ K = 39, S = 32, O = 16).**

First find the relative molecular mass of $K_2SO_4$.

$$M_r = (2 \times 39) + 32 + (4 \times 16) = 174$$

$M_r$ = **Mass of 1 mole of molecules in g**

$$\text{Moles} = \frac{\text{Mass}}{M_r}$$

$$\text{Mass} = \text{Moles} \times M_r$$

$$\text{Mass of } K_2SO_4 = 0.25 \times 174$$
$$= 43.5 \text{ g}$$

**Problem set 1.2**

1. Determine the mass of 0.02 moles of sodium chloride ($A_r$ Na = 23, Cl = 35.5).
2. Determine the mass of 0.2 mmol calcium oxide ($A_r$ Ca = 40, O = 16).

mmol = millimole

3. How many moles are present in (a) 0.254 g and (b) 6 mg of $I_2$ ($A_r$ I = 127).
4. What mass of element or compound is required to give 0.03 mole of (a) NaOH; (b) $P_4O_{10}$; (c) $S_8$; (d) $CuSO_4 \cdot 5H_2O$; (e) $Na_3PO_4$; (f) $P_4$; (g) $Cu_2O$ ($A_r$ H = 1, O = 16, Na = 23, P = 31, S = 32, Cu = 63.5).

**Worked example 1.2**

**Determine the number of atoms in 6 g of chromium ($A_r$ Cr = 52).**

First find the number of moles of chromium in 6 g of the metal:

$$\text{Moles} = \frac{\text{Mass}}{M_r}$$

$$\text{Moles of chromium} = \frac{6}{52} = 0.115$$

Metals: see H&C Chapter 7

Metallic chromium is composed of atoms.
1 mole of chromium contains
the Avogadro number of atoms = $6.022 \times 10^{23}$ atoms
0.115 moles of chromium contains $0.115 \times 6.022 \times 10^{23}$ atoms
= $6.925 \times 10^{22}$ atoms

White phosphorus:
see H&C Section 7.8

| Worked example 1.3 | **Determine the number of atoms in 6.2 g of white phosphorus ($A_r$ P = 31).** |

White phosphorus is a molecular solid containing $P_4$ molecules (Figure 1.3).
For $P_4$,   $M_r = 4 \times 31 = 124$
To find the number of moles of $P_4$ molecules in 6.2 g of white phosphorus:

$$\text{Moles} = \frac{\text{Mass in g}}{M_r} \qquad \text{Moles} = \frac{6.2}{124} = 0.05$$

1 mole of white phosphorus contains $6.022 \times 10^{23}$ $P_4$ molecules.
0.05 moles of white phosphorus contains $0.05 \times 6.022 \times 10^{23}$ molecules
$$= 3.011 \times 10^{22} \text{ molecules.}$$
Each $P_4$ molecule contains 4 atoms, and so the number of atoms in 0.05 moles of white phosphorus = $1.204 \times 10^{23}$ atoms.

**1.3** The molecular structure of $P_4$.

---

**Problem set 1.3**

Structures of elements:
see H&C Chapter 7

1. How many moles of molecules are present in (a) 0.02 g $Br_2$; (b) 63 g $H_2O$; (c) 2.2 g solid $CO_2$ (dry ice) ($A_r$ H = 1, C = 12, O = 16, Br = 80).

2. How many atoms are present in (a) 0.5 g helium; (b) 0.2 g dihydrogen; (c) $5.12 \times 10^{-4}$ g orthorhombic sulfur; (d) 6.9 g sodium ($A_r$ He = 4, H = 1, S = 32, Na = 23).

3. Radium-224 is radioactive and decays by emitting an $\alpha$-particle:
$$^{224}_{88}\text{Ra} \rightarrow {}^{220}_{86}\text{Ra} + {}^{4}_{2}\alpha$$
An $\alpha$-particle is a helium nucleus and the decay of radium produces helium gas. Rutherford and Geiger determined that radium emitted $\alpha$-particles at a rate of $7.65 \times 10^{12}$ s$^{-1}$ mol$^{-1}$, and that this produced helium at a rate of $2.9 \times 10^{-10}$ dm$^3$ s$^{-1}$. If 1 mole of helium occupies 22.7 dm$^3$, estimate a value for the Avogadro constant.

4. Estimate the mass of one atom of silicon ($A_r$ Si = 28.1).

## GAS LAWS, IDEAL GASES AND THE KINETIC THEORY OF GASES

### The kinetic theory of gases

Gas laws: see H&C
Section 1.9

The kinetic theory of gases is a model which explains the observed gas laws. In an *ideal gas,* we assume the following (the *postulates* of the kinetic theory of gases):

- the volume occupied by the particles (atoms or molecules) is negligible,
- the particles are in continuous, random motion,
- there are no attractive or repulsive interactions between the particles,
- collisions of the particles with the walls of their container result in the pressure, $P$, exerted by the gas,
- collisions between particles are *elastic* and no kinetic energy is lost upon collision,

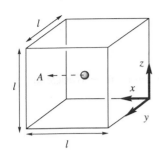

**1.4** A particle in a cubic container of side $l$ collides with a wall at point $A$; the collison is elastic and no kinetic energy is lost. After the impact, the particle rebounds and travels across the container to the opposite wall where another collision occurs. The process continues.

• the average kinetic energy, *KE,* is directly proportional to the temperature (in kelvin) of the gas.

By treating the kinetic theory of gases in a quantitative manner, we can derive the ideal gas law.

Consider a gaseous particle of mass $m$ moving in a straight-line path within a closed cubic box of side $l$. The motion continues on the same path until there is a collision with the wall of the container (Figure 1.4). The velocity, $v$, of the particle can be resolved into three directions coincident with the three Cartesian axes $(x,y,z)$. If the particle is travelling in the $x$ direction, it has a momentum of $mv_x$ (equation 1.3) and on impact, it undergoes a *change in momentum* (equation 1.4). The new velocity is $-v_x$.

$$\text{Momentum} = \text{mass} \times \text{velocity} \tag{1.3}$$

$$\text{Change in momentum} = (mv_x) - (-mv_x) = 2mv_x \tag{1.4}$$

Every time the particle travels across the container (distance $l$), a collision occurs (Figure 1.4). The number of collisions per second is given by equation 1.5 and it follows that the change in momentum per second is given by equation 1.6.

$$\text{Number of collisions per second} = \frac{\text{Velocity}}{\text{Distance}} = \frac{v_x}{l} \tag{1.5}$$

$$\text{Change in momentum per second} = \left(2mv_x\right) \times \left(\frac{v_x}{l}\right) = \frac{2mv_x^{\,2}}{l} \tag{1.6}$$

For $n$ particles in the box, travelling in all directions, the total change in momentum per second is given by equation 1.7 where $\bar{c}^2$ is the *root mean square velocity*. By Newton's second law of motion, the force exerted on the walls of the container by the impact of the particles equals the rate of change of momentum.

➤ Newton's laws of motion: see H&C Box 2.1

$$\text{Force} = \text{Total change in momentum per second} = \frac{2nm\bar{c}^2}{l} \tag{1.7}$$

*Speed* **is the rate of change of position with time; no direction is stated.** *Velocity* **is the rate of change of position in a given direction. Speed is a scalar quantity, and velocity is a vector.**

From the force, we can find the pressure since pressure is force per unit area. The total area is equal to that of the six walls of the container (equation 1.8) and the pressure is given by equation 1.9.

$$\text{Area of each wall} = l^2 \qquad \text{Total area} = 6 \times l^2 \tag{1.8}$$

$$\text{Pressure} = \frac{\text{Force}}{\text{Area}} = \frac{\left(\dfrac{2nm\bar{c}^2}{l}\right)}{6l^2} = \frac{nm\bar{c}^2}{3l^3} \tag{1.9}$$

We now have an equation that relates the pressure, $P$, to the volume, $V$, since the term $l^3$ equals the volume of the container. From equation 1.9 we can write equation 1.10, and it is possible to express the equation in terms of the kinetic energy, *KE* as is done in equation 1.11.

$$P = \frac{nm\bar{c}^2}{3V} \tag{1.10}$$

Since $\quad KE = \frac{1}{2}m\bar{c}^2 \quad$ we can write $\quad P = \frac{2n}{3V} \times (KE) \tag{1.11}$

One of the postulates of the kinetic theory of gases stated that the average kinetic energy is directly proportional to the temperature of the gas, and so equation 1.11 can be rewritten to give a relationship between pressure, volume and temperature (equation 1.12) — the $^2/_3$ disappears into the proportionality constant.

$$\frac{PV}{n} \propto T \tag{1.12}$$

This result, obtained from a *theoretical model*, has a parallel with the ideal gas law (equation 1.13) which has an *experimental* foundation.

$$PV = nRT \tag{1.13}$$

A combination of equations 1.11–1.13 allows us to relate the kinetic energy of an ideal gas to the temperature (equation 1.14).

For 1 mole of an ideal gas: $\qquad KE = \frac{3}{2}RT \tag{1.14}$

**Problem set 1.4**

**Relevant worked examples, and equations needed may be found in H&C Section 1.9**

**1 bar = $10^5$ Pa**

Data needed:   Volume of 1 mole of an ideal gas at $10^5$ Pa and 273 K = 22.7 dm$^3$
Molar gas constant, $R = 8.314$ J K$^{-1}$ mol$^{-1}$

1. What is the volume occupied by 2.5 mol $CO_2$ at 298 K at 1 bar pressure?
2. What is the change in volume when the pressure exerted on 0.5 mol NO is increased from $10^5$ Pa to $6 \times 10^5$ Pa at a constant temperature of 290 K?
3. If 50 dm$^3$ of helium is heated from 295 K to 400 K at constant pressure, what is the increase in volume?
4. What volume of $I_2$ vapour is produced at 480 K and 1 bar pressure from 2.54 g of solid $I_2$ ($A_r$ I = 127)?
5. What change in pressure is needed to increase the volume of a sample of CO, initially at 1 bar pressure, from 25 to 28 dm$^3$ at constant temperature?
6. A 66 dm$^3$ sample of gas at 300 K and $10^5$ Pa, contains 0.4 mol $H_2$ and 1.7 mol He. (a) Determine the total number of moles of gas in the mixture. (b) Are He and $H_2$ the only gases present? (c) What are the partial pressures of He and $H_2$?
7. What mass of dry ice (solid $CO_2$) would give 1.5 dm$^3$ of gaseous $CO_2$ on vaporizing at 298 K and 101.3 kPa pressure ($A_r$ C = 12, O = 16)?
8. How many moles of $SO_2$ are present in 40 cm$^3$ of the gas at 300 K and a pressure of $1.1 \times 10^5$ Pa?
9. By applying suitable changes in temperature or pressure, give *two* ways in which you could increase the volume of a sample of helium from 23 dm$^3$ to 35 dm$^3$ if the initial temperature is 290 K, and initial pressure is 1 bar.

## Graham's Law of effusion

(a)

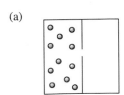

(b)

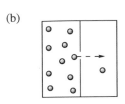

**1.5** (a) A gas is contained in a chamber which is separated from another chamber at lower pressure by a wall in which there is a pinhole.
(b) Gas molecules (or atoms) will pass from the region of higher to lower pressure until the pressures in the two containers are equal. This process is called *effusion*.

Consider the two containers shown in Figure 1.5; they are separated by a wall containing a tiny hole (a 'pinhole'). The pressure of the gas in the left-hand container is higher than that in the right-hand vessel — if there are no gas molecules at all in the right-hand container, the pressure is zero and the right-hand chamber is said to be *evacuated*. (In practice, the lowest pressures attainable experimentally are of the order of $10^{-8}$ Pa). Gas molecules (or atoms if it is a monatomic gas) will pass through the pinhole from left to right until the pressures exerted by the gases in each container are equal. This process is called *effusion*.

Graham's Law of effusion (proven experimentally) is given in equation 1.15.

$$\text{Rate of diffusion of a gas} \propto \frac{1}{\sqrt{\text{Relative molecular mass}}} \qquad (1.15)$$

The law also applies to the process of diffusion, i.e. the rate at which gases, initially held separately from each other, mix (see Figure 1.3 in H&C).

*Exercise*: Confirm the validity of Graham's Law by plotting a suitable graph using the data in Table 1.1.

**Table 1.1**   Rates of effusion for a range of gases, determined experimentally in the same piece of apparatus.

| Gaseous compound | Molecular formula | Rate of effusion / $cm^3 \ s^{-1}$ |
|---|---|---|
| Methane | $CH_4$ | 1.36 |
| Neon | Ne | 1.22 |
| Dinitrogen | $N_2$ | 1.03 |
| Carbon dioxide | $CO_2$ | 0.83 |
| Sulfur dioxide | $SO_2$ | 0.71 |
| Dichlorine | $Cl_2$ | 0.65 |

**Problem set 1.5**

1. Figure 1.6 shows a classic experiment to illustrate the dependence of the rate of diffusion on relative molecular mass. Mark on the diagram the approximate position at which ammonium chloride first forms.

2. If $V$ cm$^3$ of $NH_3$ effuse at a rate of 2.25 cm$^3$ s$^{-1}$, and $V$ cm$^3$ of an unknown gas **X** effuse at a rate of 1.4 cm$^3$ s$^{-1}$ under the same experimental conditions, what is the relative molecular mass of **X**? ($A_r$ N = 14, H = 1). Suggest a possible identity for **X**.

**1.6** A glass tube containing two cotton wool plugs soaked in aqueous $NH_3$ and HCl; the ends of the tube are sealed with stoppers. Gaseous $NH_3$ and HCl will diffuse and react to give ammonium chloride (a white solid).

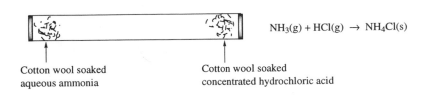

Cotton wool soaked aqueous ammonia

Cotton wool soaked concentrated hydrochloric acid

$NH_3(g) + HCl(g) \rightarrow NH_4Cl(s)$

## SOLUTION CONCENTRATION

A solution of *accurately known* concentration is prepared in a *volumetric flask* which is calibrated for a given volume. For an aqueous solution of given concentration, the correct amount of solid required should be dissolved in a volume of water in a beaker, the volume being *less* than the final volume in the flask. This solution is transferred to the volumetric flask and the volume is then *made up to the mark* on the flask.

| Worked example 1.4 |
| --- |

**What mass of sodium nitrate is required to prepare a 250 cm³ 0.1 M aqueous solution? ($A_r$ Na = 23, N = 14, O = 16).**

The formula of sodium nitrate is $NaNO_3$.
First, determine the mass of 1 mole:    $M_r = 23 + 14 + (3 \times 16) = 85$
Find how many moles are present in 250 cm³ 0.1 M solution.

$$\text{Moles} = \frac{\text{Volume (in cm}^3) \times \text{Concentration}}{1000}$$

$$\text{Number of moles} = \frac{250 \times 0.1}{1000} = 2.5 \times 10^{-2}$$

Now find the mass of $NaNO_3$ required.

$$\text{Mass} = \text{Moles} \times M_r$$
$$\text{Mass of NaNO}_3 \text{ needed} = 2.5 \times 10^{-2} \times 85$$
$$= 2.125 \text{ g}$$

*Warning!* Do *not* approximate the value if you are preparing an *accurate* solution.

---

| Problem set 1.6 |
| --- |

**Data needed:**
$A_r$ Na = 23; O = 16; N = 14; H = 1; S = 32; K = 39; I = 127; Cl = 35.5

1.  How many moles of NaCl are present in 50 cm³ 0.2 M aqueous solution.
2.  How many moles of sodium iodide are present in 25 cm³ 0.5 M solution?
3.  How many moles of aqueous hydroxide ions are present in 40 cm³ 2 M solution?
4.  Determine the concentration of 150 cm³ aqueous solution containing 0.71 g of sodium sulfate.
5.  If 8.3 g potassium iodide is dissolved in 250 cm³ water, what is the concentration of the solution? How many moles of $K^+$ ions are present in this solution?
6.  What mass of ammonium nitrate ($NH_4NO_3$) is needed to prepare 100 cm³ of a 0.25 M aqueous solution?
7.  What mass of sodium chloride is needed to prepare a 250 cm³ 0.1 M aqueous solution?
8.  A 100 cm³ 0.1 M aqueous solution of sodium iodide is diluted with water to 250 cm³. What is the concentration of the final solution?
9.  100 cm³ of a 0.2 M aqueous solution of potassium chloride is heated until the volume of the solution has been reduced to 40 cm³. What is the concentration of the final solution?

## STOICHIOMETRY AND BALANCED EQUATIONS

A balanced or *stoichiometric* equation gives the molecular ratio in which the reactants combine and the corresponding ratios of products.

**Problem set 1.7**

1. Balance the following equations by inserting appropriate numbers:

$CaH_2(s) + \quad H_2O(l) \rightarrow \quad Ca(OH)_2(aq) + \quad H_2(g)$

$Na(s) + \quad H_2O(l) \rightarrow \quad NaOH(aq) + \quad H_2(g)$

$C_6H_{14}(l) + \quad O_2(g) \rightarrow \quad CO_2(g) + \quad H_2O(l)$

$SnO_2(s) + \quad KOH(aq) + \quad H_2O(l) \rightarrow \quad K_2[Sn(OH)_6](aq)$

$Pb(NO_3)_2(aq) + 2\,NaCl(aq) \rightarrow \quad PbCl_2(s) + 2\,NaNO_3(aq)$

$MnO_2(s) + 4HCl(aq) \rightarrow \quad MnCl_2(aq) + \quad Cl_2(g) + 2\,H_2O(l)$

**Data needed:**
$A_r$ H = 1; N = 14; O = 16;
Na = 23; Cl = 35.5; Ca = 40;
Mn = 55; Pb = 207

2. Using the balanced equations from question 1, determine:
(a) the mass of sodium that is completely consumed by 0.05 mol water;
(b) the mass of $PbCl_2$ precipitated when 0.01 mol $Pb(NO_3)_2$ reacts with an excess of aqueous sodium chloride;
(c) the volume of carbon dioxide formed at $10^5$ Pa and 298 K when 0.8 mol $C_6H_{14}$ burns in dioxygen (1 mol $CO_2$ occupies 22.7 dm$^3$ at $10^5$ Pa, 273 K);
(d) the volume of $H_2$ formed at $10^5$ Pa and 295 K when 2.1 g of $CaH_2$ reacts completely with water (1 mol $H_2$ occupies 22.7 dm$^3$ at $10^5$ Pa, 273 K);
(e) the mass of $MnO_2$ needed to give 0.15 mol $Cl_2$ during a reaction with hydrochloric acid.

3. A motor vehicle safety air-bag contains $NaN_3$ and $KNO_3$. On impact, $N_2$ is produced by the successive reactions:

$2NaN_3 \rightarrow 2Na + 3N_2$

$10Na + 2KNO_3 \rightarrow K_2O + 5Na_2O + N_2$

What volume of $N_2$ is produced from 130g of $NaN_3$ at 300 K and $1 \times 10^5$ Pa pressure? ($A_r$ Na = 23, N = 14; vol. 1 mol $N_2$ = 22.7 dm$^3$ at $10^5$ Pa, 273 K).

## OXIDATION STATES AND BALANCED REDOX REACTIONS

**Worked example 1.5**

**Assign oxidation states to *each* element in $Cr_2(SO_4)_3$.**

First, decide which of the elements is likely to have a fixed oxidation state.

➤

**Guidelines for assigning oxidation states are set out in H&C Section 1.16**

O is likely to be in oxidation state −2.

S and Cr could both have variable oxidation states.

Are there any hidden groups for which you know the overall charge?

$Cr_2(SO_4)_3$ contains the sulfate ion $[SO_4]^{2-}$; the oxidation state of the sulfur atom is determined by ensuring that the sum of the oxidation states is −2 (the overall charge on the ion).

(Oxidation state of S) + (4 × Oxidation state of O) = −2

(Oxidation state of S) + (−8) = −2

Oxidation state of S = +6

The oxidation state of chromium can now be found by one of two methods.

**Method 1**

$(2 \times \text{Oxidation state of Cr}) + (3 \times \text{Oxidation state of S})$
$$+ (12 \times \text{oxidation state of O}) = 0$$
$(2 \times \text{Oxidation state of Cr}) + \{3 \times (+6)\} + \{12 \times (-2)\} = 0$
$2 \times \text{Oxidation state of Cr} = 24 - 18 = 6$
$\text{Oxidation state of Cr} = +3$

**Method 2**

The overall charge on the two chromium centres must balance the charge on the three sulfate ions, and so each chromium ion must be $Cr^{3+}$, i.e. an oxidation state of +3.

---

**Problem set 1.8**

1.  Assign oxidation states to each element in the following molecules: (a) $H_2S$; (b) $CO_2$; (c) $SOCl_2$; (d) $CaF_2$; (e) $MnO_2$; (f) $GeCl_4$; (g) $SF_6$; (h) $POCl_3$; (i) $KMnO_4$; (j) $CF_2Cl_2$; (k) $H_2O_2$ (*care!*); (l) $S_8$; (m) $C_{60}$; (n) $Fe_2O_3$.

2.  Assign oxidation states to each element in each of the following: (a) $[NO]^+$; (b) $[Cr_2O_7]^{2-}$; (c) $[SnCl_6]^{2-}$; (d) $[H_3O]^+$; (e) $[BH_4]^-$; (f) $[ClO_4]^-$; (g) $[ClO_3]^-$.

## Oxidation and reduction

**Redox =**
**Reduction–oxidation**

Oxidation may involve gaining oxygen, losing hydrogen, or losing one or more electrons. Reduction may involve gaining hydrogen, losing oxygen, or gaining one or more electrons.

**Problem set 1.9**

1.  In each of the following reactions, which species is oxidized and which is reduced. Which of the reactants is the oxidizing agent and which is the reducing agent?

    (a) $2K(s) + 2H_2O(l) \rightarrow 2KOH(aq) + H_2(g)$

    (b) $Mg(s) + 2HCl(aq) \rightarrow MgCl_2(aq) + H_2(g)$

    (c) $Fe_2O_3(s) + 2Al(s) \xrightarrow{\Delta} 2Fe(s) + Al_2O_3(s)$

    (d) $N_2(g) + 2O_2(g) \rightarrow N_2O_4(g)$

    (e) $H_2(g) + Cl_2(g) \rightarrow 2HCl(g)$

    (f) $3Cu(s) + [Cr_2O_7]^{2-}(aq) + 14H^+(aq) \rightarrow 3Cu^{2+}(aq) + 2Cr^{3+}(aq) + 7H_2O(l)$

    (g) $P_4(s) + 5O_2(g) \rightarrow P_4O_{10}(s)$

    (h) $SiCl_4(g) + 2H_2(g) \rightarrow Si(s) + 4HCl(g)$

    (i) $Cl_2(g) + H_2O_2(aq) + 2[OH]^- \rightarrow 2Cl^-(aq) + 2H_2O(l) + O_2(g)$

    (j) $[IO_3]^-(aq) + 5I^-(aq) + 6H^+(aq) \rightarrow 3I_2(aq) + 3H_2O(l)$

2.  For each of the above reactions, show that the *changes* in oxidation states for the reduction and oxidation steps balance each other.

# ENTHALPY CHANGES AND HESS'S LAW

➤ **Values of $\Delta_f H^{\circ}(298\ K)$ for selected compounds are listed in H&C Appendix 11**

The *standard enthalpy (heat) of formation* of a compound, $\Delta_f H^{\circ}(298\ K)$, is the enthalpy change that accompanies the formation of a compound in its standard state from its constituent elements, each being in its standard state.

The standard enthalpy of formation of an element *in its standard state* is zero.

A positive value of $\Delta H$ corresponds to an *endothermic* reaction (heat energy is taken in from the surroundings) and a negative value to an *exothermic reaction* (heat energy is lost to the surroundings).

The change in standard enthalpy of a reaction, $\Delta_r H^{\circ}$, is given by equation 1.16.

$$\Delta_r H^{\circ}(298\ K) = \Sigma\Delta_f H^{\circ}(298\ K)_{products} - \Sigma\Delta_f H^{\circ}(298\ K)_{reactants} \qquad (1.16)$$

---

**Worked example 1.6**

**Using data from Appendix 11, determine the enthalpy change $\Delta_r H^{\circ}(298\ K)$ for the reaction:**

$$2CO(g) + O_2(g) \rightarrow 2CO_2(g)$$

The standard enthalpy of reaction is given by the equation:

$$\Delta_r H^{\circ}(298\ K) = \Sigma\Delta_f H^{\circ}(298\ K)_{products} - \Sigma\Delta_f H^{\circ}(298\ K)_{reactants}$$

$$\Delta_r H^{\circ}(298\ K) = [2 \times \Delta_f H^{\circ}(CO_2, g, 298\ K)] - [\{2 \times \Delta_f H^{\circ}(CO, g, 298\ K)\} + \Delta_f H^{\circ}(O_2, g, 298\ K)]$$

But $O_2(g)$ is an element in its standard state: $\Delta_f H^{\circ}(O_2, g, 298\ K) = 0$.

Look up the data needed:
$\Delta_f H^{\circ}(CO_2, g, 298\ K) = -393.5\ kJ\ mol^{-1}$
$\Delta_f H^{\circ}(CO, g, 298\ K) = -110.5\ kJ\ mol^{-1}$

$$\begin{aligned}\Delta_r H^{\circ}(298\ K) &= \{2 \times \Delta_f H^{\circ}(CO_2, g, 298\ K)\} - \{2 \times \Delta_f H^{\circ}(CO, g, 298\ K)\} \\ &= \{2 \times (-393.5)\} - \{2 \times (-110.5)\} \\ &= -566\ kJ\ per\ mole\ of\ reaction\end{aligned}$$

➤ **Caution!**

The answer is in terms of an *enthalpy change per mole of the reaction as written in the equation*. If we had simply written $-566\ kJ\ mol^{-1}$, it would have been ambiguous — did we mean $\Delta_r H^{\circ}(298\ K) = -566\ kJ$ per mole of CO or per 2 moles of CO ?

---

**Problem set 1.10**

**Data needed: see H&C Appendix 11**

1. Determine the standard enthalpy change for each of the following reactions.

(a) $2Cr(s) + 3Cl_2 \rightarrow 2CrCl_3(s)$

(b) $2H_2O_2(l) \rightarrow O_2(g) + 2H_2O(l)$

(c) $P_4O_6(s) + 2O_2(g) \rightarrow P_4O_{10}(s)$

(d) $Ca(OH)_2(s) + CO_2(g) \rightarrow CaCO_3(s) + H_2O(l)$

(e) $Fe_2O_3(s) + 2Al(s) \rightarrow 2Fe(s) + Al_2O_3(s)$

(f) $H_2(g) + Cl_2(g) \rightarrow 2HCl(g)$

(g) $CH_3CH_2OH(l) + 3O_2(g) \rightarrow 2CO_2(g) + 3H_2O(l)$

(h) $CH_2=CH_2(g) + Br_2(g) \rightarrow CH_2BrCH_2Br(l)$

➤ **Combustion data are readily measured experimentally and are a means of obtaining values of the standard heats of formation of compounds**

2.    The standard enthalpy of combustion, $\Delta_c H^\circ$(298 K) of acetone, $CH_3C(O)CH_3$, is $-1790$ kJ mol$^{-1}$. Determine the standard enthalpy of formation (298 K) of acetone.

3.    The Hess cycle below describes enthalpy changes for the addition of $H_2$ to acetylene and ethene. (a) Determine values of $\Delta H(1)$, $\Delta H(2)$ and $\Delta H(3)$ from appropriate standard enthalpies of formation. (b) Use these results to confirm the validity of Hess's Law.

$$HC{\equiv}CH(g) + 2H_2(g) \xrightarrow{\Delta H(1)} H_2C=CH_2(g) + H_2(g)$$

$$\Delta H(3) \searrow \qquad \swarrow \Delta H(2)$$

$$C_2H_6(g)$$

## EQUILIBRIA

### Le Chatelier's Principle

Le Chatelier's Principle states that when an external change is made to a system in equilibrium, the system will change in order to compensate for the change.

| Worked example 1.7 |

**What will be the effect of increasing the pressure on the following equilbrium ?**
$$2CO(g) + O_2(g) \rightleftharpoons 2CO_2(g)$$

The forward reaction involves 3 moles of gaseous reactants going to 2 moles of gaseous product.

   If the external pressure is increased, the system will oppose the change and the forward reaction will be favoured, because as the number of moles decreases, the pressure exerted by the gases decreases.

| Problem set 1.11 |

1.    The value of $\Delta_f H^\circ$($NH_3$, g, 298 K) is $-45.9$ kJ mol$^{-1}$. What is the effect on the following equilibrium of raising the temperature at constant pressure? What is the effect of increasing the pressure at constant temperature?

$$N_2(g) + 3H_2(g) \rightleftharpoons 2NH_3(g)$$

2.    The value of $\Delta_f H^\circ$($NO_2$, g, 298 K) = $+34.2$ kJ mol$^{-1}$. What is the effect on the following equilibrium of raising the temperature at constant pressure? What is the effect of increasing the pressure at constant temperature?

$$N_2(g) + 2O_2(g) \rightleftharpoons 2NO_2(g)$$

3.    What is the effect of adding aqueous alkali to the following equilibrium?

$$[Cr_2O_7]^{2-}(aq) + 2[OH]^-(aq) \rightleftharpoons 2[CrO_4]^{2-}(aq) + H_2O(l)$$

4.    What is the effect on the following equilibrium of (a) adding propanoic acid, and (b) removing benzyl propanoate by distillation?

$$CH_3CH_2CO_2H + PhCH_2OH \underset{}{\overset{H^+}{\rightleftharpoons}} CH_3CH_2C\overset{O}{\underset{OCH_2Ph}{\diagdown}} + H_2O$$

Propanoic acid       Benzyl alcohol       Benzyl propanoate

## Solution equilibrium constant, $K_c$

The equilibrium constant, $K_c$, for a solution system:

$$aA + bB \rightleftharpoons cC + dD$$

is given by equation 1.17 where [A] is the concentration of component A, [B] is the concentration of component B, etc.

$$K_c = \frac{[C]^c[D]^d}{[A]^a[B]^b} \tag{1.17}$$

**Worked example 1.8**    **In aqueous solution, methylamine acts as a weak base:**

$$CH_3NH_2(aq) + H_2O(l) \rightleftharpoons [CH_3NH_3]^+(aq) + [OH]^-(aq)$$

**Determine the equilibrium constant if the initial concentration of methylamine is 0.1 mol dm⁻³, and at equilibrium the concentration of hydroxide ions is $6.6 \times 10^{-3}$ mol dm⁻³.**

Let the total volume be 1 dm³ :

$$CH_3NH_2(aq) + H_2O(l) \rightleftharpoons [CH_3NH_3]^+(aq) + [OH]^-(aq)$$

Moles initially:        0.1        excess        0        0
Moles at equilm:   0.1−(6.6 × 10⁻³)   excess   6.6 × 10⁻³   6.6 × 10⁻³

$$K_c = \frac{[C_6H_5NH_3^+][OH^-]}{[C_6H_5NH_2][H_2O]}$$

➤ But since [H₂O] is taken to be unity:     $$K_c = \frac{[C_6H_5NH_3^+][OH^-]}{[C_6H_5NH_2]}$$
**See H&C: Section 11.9**

$$K_c = \frac{\left(\dfrac{6.6\times10^{-3}}{V}\right)\left(\dfrac{6.6\times10^{-3}}{V}\right)}{\left(\dfrac{0.1-\left(6.6\times10^{-3}\right)}{V}\right)}$$

The volume $V$ is 1 dm³ and the expression for $K_c$ is:

$$K_c = \frac{\left(6.6 \times 10^{-3}\right)^2}{0.1 - \left(6.6 \times 10^{-3}\right)}$$

$$= 4.7 \times 10^{-4} \text{ mol dm}^{-3}$$

➤ **See H&C: Section 11.9**

For a weak base (or acid) HX, it is common to use the approximation:

$$[HX]_{\text{equilm}} \approx [HX]_{\text{initial}}$$

Test how valid this approximation is for methylamine by recalculating $K_c$ using a value of $[CH_3NH_2]_{\text{equilm}} \approx 0.1$ mol dm⁻³.

---

**Problem set 1.12**

1. Acetic acid dissociates in aqueous solution according to the equilibrium:

$$CH_3CO_2H(aq) + H_2O(l) \rightleftharpoons [CH_3CO_2]^-(aq) + [H_3O]^+(aq)$$

If the initial concentration of acetic acid is 0.4 mol dm⁻³, and at equilibrium, the concentration of $[CH_3CO_2]^-$ ions is $2.6 \times 10^{-3}$ mol dm⁻³, what is the value of equilibrium constant?

2. Diiodine is very sparingly soluble in water, and laboratory solutions are usually made up in aqueous potassium iodide in which $I_2$ is soluble giving salts of the $[I_3]^-$ ion (Figure 1.7):

$$I_2(aq) + I^-(aq) \rightleftharpoons [I_3]^-(aq)$$

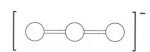

**1.7** The linear structure of the $[I_3]^-$ anion.

When 2.54 g of $I_2$ are added to 1 dm³ 0.5 M aqueous solution of potassium iodide, and the solution is allowed to reach equilibrium, it is found that the solution contains $9.8 \times 10^{-3}$ moles of $[I_3]^-$ ion. What is the equilibrium constant? (Assume that the volume of solution is 1 dm³ after the addition of the $I_2$; $A_r$ I = 127).

3. Acetic acid and ethanol react to give an ester and water:

$$\underset{\text{Acetic acid}}{CH_3CO_2H(l)} + \underset{\text{Ethanol}}{CH_3CH_2OH(l)} \rightleftharpoons CH_3CO_2C_2H_5(l) + H_2O(l)$$

When 1 mole of acetic acid and 0.5 mole of ethanol are combined and the reaction mixture is allowed to reach equilibrium at 298 K, 0.425 moles of $CH_3CO_2C_2H_5$ are present. What is the equilibrium constant?

➤ **As this is an acid dissociation constant, we would usually write the equilibrium constant as $K_a$: see Chapter 11**

4. The value of $K_c$ for the following equilibrium is 0.01 mol dm⁻³.

$$H_3PO_3(aq) + H_2O(l) \rightleftharpoons [H_3O]^+(aq) + [H_2PO_3]^-(aq)$$

If an aqueous 0.1 M solution of $H_3PO_3$ is allowed to reach equilibrium, what is the concentration of $[H_2PO_3]^-$ ions present in solution?

### Gaseous equilibria: the equilibrium constant, $K_p$

The equilibrium constant, $K_p$, for a gaseous system:

$$aA(g) + bB(g) \rightleftharpoons cC(g) + dD(g)$$

is given by equation 1.18 where $P_A$ is the partial pressure of component A, $P_B$ is the partial pressure of component B, etc. The partial pressure of a gas in a mixture of gases is given by equation 1.19.

$$K_p = \frac{(P_C)^c (P_D)^d}{(P_A)^a (P_B)^b} \tag{1.18}$$

$$\text{Partial pressure of gas X} = \left( \frac{\text{Moles of X}}{\text{Total moles of gas}} \right) \times \text{Total pressure} \tag{1.19}$$

---

| Worked example 1.9 |
|---|

**At 350 K, the value of $K_p$ for the dissociation of gaseous $N_2O_4$ is 3.9 bar.**

$$N_2O_4(g) \rightleftharpoons 2NO_2(g)$$

**If 0.3 moles of $N_2O_4$ are allowed to reach equilibrium with $NO_2$ under a total pressure of 2 bar, how many moles of $N_2O_4$ remain?**

➤ 
**1 bar = $10^5$ Pa
You can work in either set of units, *but you must be consistent***

Find the number of moles of each component in the equilibrium mixture and then find their partial pressures.

Let $x$ moles of $N_2O_4$ produce $2x$ moles of $NO_2$.

$$N_2O_4(g) \rightleftharpoons 2NO_2(g)$$

Initial moles:          0.3          0
Moles at equilm:    $(0.3–x)$      $2x$

Total moles of gas at equilibrium = $(0.3–x) + 2x = 0.3 + x$

The total pressure is 2 bar. Partial pressures at equilibrium:

$$P_{N_2O_4} = \frac{(0.3-x)}{(0.3+x)} \times 2$$

$$P_{NO_2} = \frac{2x}{(0.3+x)} \times 2 = \frac{4x}{(0.3+x)}$$

The equilibrium constant is given by:

$$K_p = \frac{(P_{NO_2})^2}{(P_{N_2O_4})}$$

$$K_p = 3.9 = \frac{\left\{\dfrac{4x}{(0.3+x)}\right\}^2}{\left\{\dfrac{(0.3-x)}{(0.3+x)}\right\}\times 2}$$

$$= \frac{\left(16x^2\right)\times(0.3+x)}{(0.3+x)^2 \times(0.3-x)\times 2}$$

$$= \frac{8x^2}{0.09 - x^2}$$

$$3.9\times\left(0.09 - x^2\right) = 8x^2$$

$$x = 0.17$$

The number of moles of $N_2O_4$ remaining $= 0.3 - x = 0.13$

---

**Problem set 1.13**

1.  Dinitrogen and dihydrogen react to give ammonia:

    $$N_2(g) + 3H_2(g) \rightleftharpoons 2NH_3(g)$$

    If 0.5 moles of $N_2$ and 2 moles of $H_2$ are combined at 400 K and 1 bar pressure and the system allowed to reach equilibrium, 0.81 moles of $NH_3$ are present in the equilibrium mixture. Determine the value of $K_p$ under these conditions.

➤ **Sulfuric acid manufacture: see H&C Section 13.10**

2.  The oxidation of $SO_2$ is a stage in the manufacture of sulfuric acid.

    $$2SO_2(g) + O_2(g) \rightleftharpoons 2SO_3(g)$$

    If 2 moles of $SO_2$ react with 0.5 moles of $O_2$ at 1100 K and 1 bar pressure and the system is left to reach equilibrium, the final mixture contains 0.24 moles of $SO_3$. What is the value of $K_p$ under these conditions?

3.  The formation of HI from its constituent elements takes place according to the reaction:

    $$H_2(g) + I_2(g) \rightleftharpoons 2HI(g)$$

    At 500 K, the value of $K_p = 160$. If 1 mole of $H_2$ and 1 mole of $I_2$ are combined, how many moles of HI are present when the system has reached equilibrium if the total pressure is 4 bar?

4.  Consider the equilibrium:

    $$H_2(g) + CO_2(g) \rightleftharpoons H_2O(g) + CO(g)$$

    At 800 K, the value of $K_p$ is 0.29. If 0.8 moles of $H_2$ and 0.6 moles of $CO_2$ react under a pressure of 1 bar, how many moles of $CO_2$ will remain when the reaction mixture has reached equilibrium?

# 2 Atoms and atomic structure

> **Topics**
> - Quantum numbers and orbital types
> - Atomic orbitals and radial distribution functions
> - Penetration and shielding
> - The atomic spectrum of hydrogen and selection rules
> - The *aufbau* principle and ground state electronic configurations
> - The octet rule
> - Periodicity

## QUANTUM NUMBERS AND ORBITAL TYPES

An atomic orbital is defined by a unique set of three quantum numbers:

$$n, l \text{ and } m_l$$

and an electron in an atom is defined by a unique set of four quantum numbers:

$$n, l, m_l \text{ and } m_s.$$

$n$    Principal quantum number with values of $n = 1, 2, 3, 4, 5 \dots \infty$.

$l$    Orbital quantum number with values of $l = 0, 1, 2, 3, 4 \dots (n-1)$.

$m_l$    Magnetic quantum number with values of $-l, (-l+1) \dots 0 \dots (l-1), l$.

$m_s$    Magnetic spin quantum number with values of $+\frac{1}{2}$ or $-\frac{1}{2}$.

| Orbital type | $s$ | $p$ | $d$ | $f$ |
|---|---|---|---|---|
| Value of $l$ | 0 | 1 | 2 | 3 |

---

**Worked example 2.1**

**What sets of quantum numbers describe the five 3*d* atomic orbitals?**

For the $3d$ AOs, the principal quantum number $n = 3$.

For $n = 3$, the values of $l$ are 0, 1, 2.

Values of $n = 3$, $l = 0$ and $n = 3$, $l = 1$ correspond to the $3s$ and $3p$ AOs.

Each $3d$ atomic orbital has a value of $l = 2$.

For $l = 2$, values of $m_l = -2, -1, 0, +1, +2$, meaning that there are five $3d$ AOs.

The sets of quantum numbers which describe these orbitals are:

$$n = 3, l = 2, m_l = -2$$
$$n = 3, l = 2, m_l = -1$$
$$n = 3, l = 2, m_l = 0$$
$$n = 3, l = 2, m_l = +1$$
$$n = 3, l = 2, m_l = +2$$

1.  To which orbital types do the following values of $l$ correspond (a) 1; (b) 3; (c) 2; (d) 0 ?
2.  How many atomic orbitals are there in the $5f$ sub-shell?
3.  How many orbitals are there in the shell with $n = 4$?
4.  What sets of quantum numbers describe the $2p$ atomic orbitals?
5.  How many sub-shells are present for orbitals of principal quantum numbers (a) 1; (b) 2; (c) 3; (d) 4?
6.  Which sub-shells *cannot* exist in an atom (a) $1p$; (b) $2s$; (c) $3f$; (d) $5s$; (e) $2d$ ?
7.  Which quantum number tells you that two electrons in an atomic orbital possess opposite spins?
8.  Write down the sets of quantum numbers that describe the electrons in a fully occupied set of $4p$ atomic orbitals.
9.  Which atomic orbital has the set of quantum numbers $n = 3$, $l = 0$, $m_l = 0$ ? How do you distinguish between the two electrons that may occupy this orbital?

## ATOMIC ORBITALS AND RADIAL DISTRIBUTION FUNCTIONS

### Shapes of *s* and *p* atomic orbitals

The shape of an atomic orbital depends on the quantum number $l$. Figure 2.1 summarizes the spatial properties of $s$ ($l = 0$) and $p$ ($l = 1$) atomic orbitals.

1.  What do the terms *singly degenerate* and *triply (three-fold) degenerate* mean?
2.  What is a *nodal plane* ?
3.  How many nodal planes does each of the following atomic orbitals possess? (a) $s$; (b) $p$.
4.  For a given atom, list the following atomic orbitals in order of *increasing* size: $1s$, $2s$, $3s$, $4s$, $5s$. Which of these AOs is the most diffuse?
5.  For a given atom, which of the following atomic orbitals is (a) the most diffuse, (b) the least diffuse: $2p$, $4p$, $5p$?
6.  For a given atom, list the following atomic orbitals in order of *increasing* energy: $1s$, $2s$, $3s$, $2p$, $3p$.

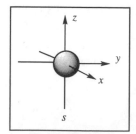

Singly degenerate

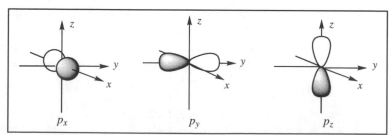

Triply degenerate

**2.1**  An $s$ atomic orbital is spherically symmetric and the boundary surface of the orbital has a constant phase. In a $p$ orbital, there is one phase change with respect to the surface boundary of the orbital; each part of the orbital is called a lobe.

### Radial distribution functions, $4\pi r^2 R(r)^2$

➤

**R(r) is the radial part of the wavefunction: see H&C Section 2.8**

Figure 2.1 showed *surface boundaries* for *s* and *p* atomic orbitals. The surface boundaries of *ns* atomic orbitals are similar, as are those for *np* AOs. The *radial distribution functions* $4\pi r^2 R(r)^2$ for *s* atomic orbitals are dependent on the principal quantum number *n*. Similarly, the function $4\pi r^2 R(r)^2$ is different for 2*p*, 3*p*, 4*p*..., 3*d*, 4*d*... and 4*f*, 5*f*... AOs. Radial distribution functions are conveniently represented in graphical form and Figure 2.2 shows plots of $4\pi r^2 R(r)^2$ as a function of *r* (the distance from the nucleus — *r* = 0 corresponds to the nucleus) for some of the orbitals of the hydrogen atom. The functions tell us the probability of finding an electron at a particular distance from the nucleus. By comparing the radial distribution functions for the 1*s*, 2*s* and 3*s*, or the 2*p* and 3*p* AOs in Figure 2.2, we see that for a given orbital type (*s* or *p*):

• the orbital extends further from the nucleus as *n* increases,
• the number of *radial nodes* increases as *n* increases.

| Atomic orbital | 1s | 2s | 3s | 4s | 5s | 6s |
|---|---|---|---|---|---|---|
| Number of radial nodes | 0 | 1 | 2 | 3 | 4 | 5 |

| Atomic orbital | 2p | 3p | 4p | 5p | 6p |
|---|---|---|---|---|---|
| Number of radial nodes | 0 | 1 | 2 | 3 | 4 |

| Atomic orbital | 3d | 4d | 5d | 6d |
|---|---|---|---|---|
| Number of radial nodes | 0 | 1 | 2 | 3 |

**Problem set 2.3**

**The electron only occupies the 1s AO in the electronic *ground state* of hydrogen. Promotion of the electron to higher energy levels produces *excited states*.**

1. What do you understand by the term *radial node*?
2. Why can the plots in Figure 2.2 never have negative values of $4\pi r^2 R(r)^2$?
3. Sketch the radial distribution functions as a function of the distance *r* from the nucleus for a (a) 3*d*, (b) 4*p*, (c) 4*s* atomic orbital of hydrogen.
4. Describe where an electron would spend its time in a 3*s* atomic orbital of a hydrogen atom.

**2.2** Radial distribution functions for 1*s*, 2*s*, 3*s*, 2*p* and 3*p* atomic orbitals of the hydrogen atom. The functions for the *s* orbitals are drawn in continuous lines and the dashed lines refer to the functions for the *p* orbitals.

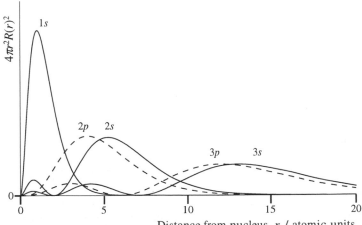

So far in this section we have considered the orbitals of a hydrogen atom. In the ground state, the single electron occupies the 1s AO but it may be excited to higher energy orbitals. In problem set 2.4 we consider 'hydrogen-like' species.

**Problem set 2.4**

1. Which of the following are hydrogen-like species (a) $H^+$; (b) $H^-$; (c) He; (d) $He^+$; (e) $Li^+$; (f) $Li^{2+}$ ?
2. How does an increase in nuclear charge affect the energy of an orbital?
3. How does an increase in nuclear charge affect the size of an orbital?
4. Figure 2.3 shows the radial distribution function for the 1s AO of hydrogen. Using the same scale (extended as necessary), sketch the approximate curve for the radial distribution function of the 1s AO of $He^+$.

**2.3** Radial distribution function for the 1s atomic orbital of hydrogen.

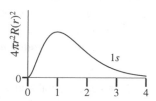

Distance from nucleus, $r$ / atomic units

## PENETRATION AND SHIELDING

The following problems are concerned with Figure 2.4.

**Problem set 2.5**

1. Using the same axis set (extended as necessary), sketch the approximate curve for the radial distribution function of the 1s atomic orbital of lithium.
2. 'The 2s AO of lithium penetrates the 1s AO.' What does this statement mean?
3. Which is more penetrating – the 2s or 2p AO of lithium?
4. The atomic number of lithium is 3. Two electrons occupy the 1s AO. Is it energetically favourable for the third electron to occupy the 2s or 2p AO?
5. Sketch an energy level diagram showing the approximate relative energies of the 1s, 2s and 2p AOs in lithium. How does this diagram differ from that of the hydrogen atom?
6. In the ground state of a boron atom ($Z = 5$), two electrons occupy the 1s AO, two the 2s AO and one the 2p AO. Assuming that the radial distribution functions for the 2s and 2p atomic orbitals in boron can be represented by a similar diagram to Figure 2.4, explain what is meant by the statement: 'the 2s electrons in boron shield the 2p electron.'
7. What is the difference between the absolute nuclear charge of an atom and the effective nuclear charge? Are these variable or constant quantities?

**2.4** Representations of the radial distribution functions of the 2s and 2p atomic orbitals of lithium.

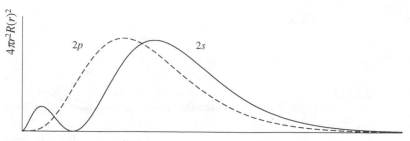

Distance from the nucleus, $r$

## THE ATOMIC SPECTRUM OF HYDROGEN AND SELECTION RULES

In the emission spectrum of atomic hydrogen, single lines are observed. Each line corresponds to a *discrete electronic transition* as an electron falls from an excited state to a lower energy state.

**Problem set 2.6**

1. Write down the selection rules concerning changes in the quantum numbers $n$ and $l$ that govern which transitions are allowed in an electronic spectrum.
2. Why does the $2p \rightarrow 2s$ transition not give rise to a spectral line in the emission spectrum of atomic hydrogen?
3. Identify each of the following $n' \rightarrow n$ transitions as being in either the Balmer or Lyman series (a) $2 \rightarrow 1$; (b) $3 \rightarrow 2$; (c) $5 \rightarrow 2$; (b) $4 \rightarrow 1$.
4. The frequencies of some consecutive spectral lines in the Lyman series of atomic hydrogen are 2.466, 2.923, 3.083, 3.157 and $3.197 \times 10^{15}$ Hz. Use these values to draw a schematic representation of this part of the emission spectrum. How does the spectrum you have drawn relate to the representation of the transitions shown in Figure 2.5? Assign each of the lines in the spectrum to a particular transition.
5. What is the energy of the transition corresponding to the spectral line of frequency $3.197 \times 10^{15}$ Hz? (Planck constant $= 6.626 \times 10^{-34}$ J s)
6. On Figure 2.5, draw four transitions that form part of the Balmer series.
7. What is the difference between an absorption and an emission spectrum?
8. Explain in terms of Figure 2.5 what happens when a hydrogen atom is ionized. Write an equation to represent this process.

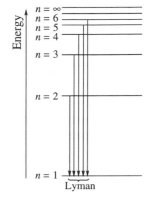

**2.5** Part of the Lyman series of transitions in the emission spectrum of atomic hydrogen.

### The Rydberg constant

The frequencies of the lines (arising from $n' \rightarrow n$ transitions) in the Lyman, Balmer or other series can be calculated using equation 2.1 where $R$ is the Rydberg constant.

*R* is used for both the Rydberg constant and the molar gas constant.

$$\text{Frequency, } v = R \times \left( \frac{1}{n^2} - \frac{1}{n'^2} \right) \tag{2.1}$$

$$R = 1.097 \times 10^7 \text{ m}^{-1} = 1.097 \times 10^5 \text{ cm}^{-1} = 3.289 \times 10^{15} \text{ Hz}$$

**Problem set 2.7**

**Electromagnetic spectrum: See H&C Appendix 4**

1. Using the frequencies in question 4 (problem set 2.6), and assignments you made, plot a suitable graph to estimate a value for the Rydberg constant.
2. From equation 2.1, determine the frequencies of the following $n' \rightarrow n$ transitions which make up part of the Paschen series in the atomic spectrum of hydrogen (a) $7 \rightarrow 3$; (b) $6 \rightarrow 3$; (c) $5 \rightarrow 3$; (d) $4 \rightarrow 3$. Determine the energy of each transition and state in which region of the electromagnetic spectrum these spectral lines appear.
3. Do the Lyman and Balmer series of lines appear in the same region of the electromagnetic spectrum as (a) the Paschen series, and (b) as each other?

## THE *AUFBAU* PRINCIPLE AND GROUND STATE ELECTRONIC CONFIGURATIONS

The *Pauli exclusion principle* states that no two electrons in an atom can have the same set of $n$, $l$, $m_l$ and $m_s$ quantum numbers.

*Hund's rule* states that when filling a degenerate set of orbitals, pairing of electrons cannot begin until *each* orbital in the set contains one electron. Electrons singly occupying orbitals in a degenerate set have parallel spins.

The *aufbau principle* combines Hund's rules and the Pauli exclusion principle with the following additional facts:

• orbitals are filled in order of increasing energy

• an orbital is fully occupied when it contains two electrons.

The order of occupying atomic orbitals in the ground state of an atom *usually* follows the sequence (lowest energy first):

$$1s < 2s < 2p < 3s < 3p < 4s < 3d < 4p < 5s < 4d < 5p < 6s < 5d \approx 4f < 6p < 7s < 6d \approx 5f$$

| Worked example 2.2 |
| --- |

**Determine the arrangement of the electrons in an atom of silicon in its ground state ($Z$ Si = 14).**

The atomic number of silicon is 14, and so there are 14 electrons.

The lowest energy orbital is the $1s$ ($n = 1$; $l = 0$; $m_l = 0$), and the next lowest AO is the $2s$ ($n = 2$; $l = 0$; $m_l = 0$). Each of the $1s$ and $2s$ orbitals is fully occupied when it contains two electrons which are spin-paired (i.e. $m_s = +^1/_2$ for one electron and $m_s = -^1/_2$ for the other electron).

The next lowest energy orbitals are the three in the degenerate set of $2p$ orbitals ($n = 2$; $l = 1$, $m_l = +1, 0, -1$). Each $2p$ AO can accommodate two electrons.

Ten electrons have so far been accommodated.

The next lowest energy orbital is the $3s$ ($n = 3$; $l = 0$; $m_l = 0$) and two electrons can be placed here.

Two electrons remain and these occupy the $3p$ level so as to obey Hund's rule.

The arrangement of electrons in a silicon atom in its ground state is represented by the diagram:

or can be written in the form:

$$1s^2 2s^2 2p^6 3s^2 3p^2 \quad \text{or} \quad [\text{Ne}]3s^2 3p^2$$

| Worked example 2.3 | **Write down the ground state electronic configuration for an $O^{2-}$ ion ($Z$ O = 8).** **Justify the answer.** |
|---|---|

The neutral oxygen atom has eight electrons.

Therefore the $O^{2-}$ anion has ten electrons.

The sequence of atomic orbitals to be filled is (in order of increasing energy):

$$1s < 2s < 2p$$

The maximum occupancy of an $s$ level is two electrons.

The maximum occupancy of the $p$ level is six electrons.

The ground state electronic configuration for the oxide ion $O^{2-}$ is:

$1s^2 2s^2 2p^6$     equivalent to     $[He]2s^2 2p^6$     equivalent to     $[Ne]$.

---

**Problem set 2.8**

1.  Write down the ground state electronic configuration of each of the following atoms: (a) N ($Z = 7$); (b) Be ($Z = 4$); (c) Cl ($Z = 17$); (d) Al ($Z = 13$); (e) K ($Z = 19$).

2.  Write down the ground state electronic configuration of each of the following ions: (a) $F^-$ ($Z = 9$); (b) $N^{3-}$ ($Z = 7$); (c) $Ca^{2+}$ ($Z = 20$); (d) $Li^+$ ($Z = 3$). What feature do these configurations have in common?

3.  Draw energy level diagrams to show the arrangement of the electrons in the ground states of (a) He ($Z = 2$); (b) B ($Z = 5$); (c) Mg ($Z = 12$); (d) O ($Z = 8$).

4.  In the ground state of an oxygen atom, which electrons would you label as *core* and which as *valence*?

5.  How many valence electrons does a fluorine atom possess?

6.  How many valence electrons does each member of group 1 possess? Write down a general electronic configuration that represents the ground state of a group 1 atom.

7.  What feature do the electronic ground states of the group 17 atoms have in common?

8.  What feature do the electronic ground states of the group 14 atoms have in common?

## THE OCTET RULE

**Problem set 2.9**

1.  What do you understand by the *octet rule*? What are its limitations?

2.  Do the following atoms tend to lose or gain electrons (a) Na; (b) F; (c) Al; (d) O; (e) Mg?

3.  Write down the ions that the following atoms are *most* likely to form (a) Na; (b) O; (c) Cl; (d) Mg; (e) N. How do you justify your choices?

4.  How are the following atoms *most* likely to achieve an octet when forming compounds (a) C; (b) N; (c) F; (d) Na; (e) Cl.

5.  Sulfur forms the halides $SCl_2$, $SF_4$ and $SF_6$. Which of these compound(s) is (are) predicted by the octet rule?

6.  Chlorine forms the series of fluorides $ClF$, $ClF_3$ and $ClF_5$. In which of these compounds is the chlorine atom obeying the octet rule?

7.    Assuming that the octet rule is obeyed, what would be the formula of a compound formed between (a) sodium and nitrogen; (b) sodium and fluorine; (c) magnesium and chlorine; (d) nitrogen and fluorine; (e) oxygen and fluorine.

## PERIODICITY

**Problem set 2.10**

1.    What feature of their ground state electronic structures do the noble gases have in common?
2.    To which group of the periodic table does an element with a ground state valence electronic configuration of $ns^2np^5$ belong?
3.    To which group of the periodic table does an element with a ground state valence electronic configuration of $ns^2np^2$ belong?
4.    Some groups of elements have collective names. Which group number is associated with the following names (a) alkaline earth metals; (b) chalcogens; (c) noble gases; (d) alkali metals; (e) halogens?
5.    Write down the general notation for the ground state valence electronic configuration of an element in group 16.
6.    Write down the general notation for the ground state valence electronic configuration of an alkali metal.
7.    Write down the general notation for the ground state valence electronic configuration of an alkaline earth metal.
8.    Write down the general notation for the ground state valence electronic configuration of an element in group 13.

**Problem set 2.11**

Using data from Table 2.6 and Appendix 11 in H&C, plot graphs on the *same* axis set to show periodic trends in boiling points and melting points for the first twenty elements ($Z = 1$–20). For carbon, mp = 3820, bp (sublimation) = 5100 K.
Using the graph, answer the following questions.

1.    Why is no melting point for helium available?
2.    Which elements have short liquid ranges?
3.    Refine your answer to question 2 to show how the noble gases differ from the other elements you have written down.
4.    In which two groups of the periodic table do you find elements with the largest liquid ranges?
5.    How do you account for the dramatic fall in melting and boiling points between elements with $Z = 6$ and 7?
6.    How do you account for the rise in melting and boiling points between elements with $Z = 2$ and 3, $Z = 10$ and 11, and $Z = 18$ and 19?

**Homonuclear covalent bonds**

---

### Topics

- Covalent radii
- Standard enthalpies of atomization and bond dissociation enthalpies
- Bond enthalpy transferability
- Lewis structures
- Valence bond theory for homonuclear diatomic molecules
- Molecular orbital theory for homonuclear diatomic molecules

---

### COVALENT RADII

**Problem set 3.1**

$r_v$ = van der Waals radius
$r_{cov}$ = covalent radius

1.  Figure 3.1 shows the trend in values of $r_v$ and $r_{cov}$ for elements from $Z = 5$ to 9.
    (a) In which part of the periodic table are these elements?
    (b) Why is the value of $r_{cov}$ smaller than that of $r_v$ for each element?
    (c) Rationalize the trend in values of $r_v$.

2.  For sulfur, $r_v = 185$ pm and $r_{cov} = 103$ pm. How may these values be estimated experimentally?

3.  Using the data in Appendix 6 of H&C, estimate the bond lengths in the gas phase diatomic molecules $F_2$, $Cl_2$, $Br_2$ and $I_2$. Comment on the trend in bond distances in going from $F_2$ to $I_2$.

**3.1** Trends in van der Waals and covalent radii for elements with atomic numbers 5 to 9.

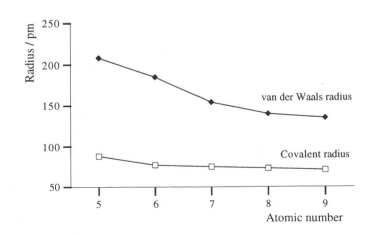

## STANDARD ENTHALPIES OF ATOMIZATION AND BOND DISSOCIATION ENTHALPIES

**Problem set 3.2**

1.  Distinguish between the use of the symbols $D$ and $\overline{D}$ for a bond dissociation enthalpy.

2.  What is the relationship between the bond dissociation enthalpy and the standard enthalpy of atomization of a gaseous *diatomic* molecule? Why can we not necessarily use similar relationships for larger homonuclear molecules?

3.  The standard enthalpies of atomization of hydrogen, chlorine, nitrogen, sulfur and phosphorus are 218, 121, 473, 277 and 315 kJ mol$^{-1}$ respectively. Determine values for the bond dissociation enthalpies of the homonuclear bonds in the molecules (a) $H_2$, (b) $Cl_2$, (c) $N_2$, (d) $S_8$ and (e) $P_4$. The structures of $P_4$ and $S_8$ are shown in Figure 3.2.

4.  Use the value of $\overline{D}$(S–S) determined in problem 3 to estimate a value for $\Delta_a H$ per mole of molecules of $S_6$ (Figure 3.2). Write an equation to define the process to which this value refers.

5.  The bond dissociation enthalpies for the *homonuclear* bonds in $C_2H_2$, $C_2H_4$, $C_2H_6$, $O_2$, $H_2O_2$, $N_2$, $N_2H_4$, $S_2H_2$ and $S_8$ are 813, 598, 346, 498, 213, 945, 275, 273 and 277 kJ mol$^{-1}$ respectively. Draw the structure of each molecule and rationalize the variations among the enthalpy values.

**3.2** The molecular structures of $P_4$, $S_8$ and $S_6$.

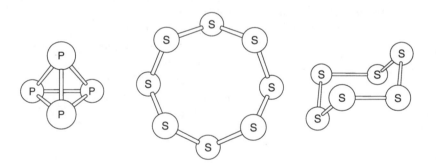

## BOND ENTHALPY TRANSFERABILITY

**Worked example 3.1**

**Using data from Appendices 10 and 11 in H&C, estimate a value for $D$(C=C).**

Ethene $H_2C=CH_2$ contains one carbon–carbon double bond and four carbon–hydrogen single bonds. Methane $CH_4$ contains only four carbon–hydrogen single bonds. We can apply the method of *bond enthalpy transferability* between methane and ethene to estimate the bond dissociation enthalpy of the C=C bond.

The data required from Appendix 10 are the standard enthalpies of atomization of carbon and hydrogen: $\Delta_a H^\circ$(C) = 717 kJ mol$^{-1}$; $\Delta_a H^\circ$(H) = 218 kJ mol$^{-1}$.

Appendix 11 lists values of the standard enthalpies of formation of methane and ethene: $\Delta_f H^\circ$(CH$_4$, g) = −74 kJ mol$^{-1}$; $\Delta_a H^\circ$(C$_2$H$_4$, g) = +52.5 kJ mol$^{-1}$.

Set up a Hess cycle to find a value of $\overline{D}$(C–H) in methane.

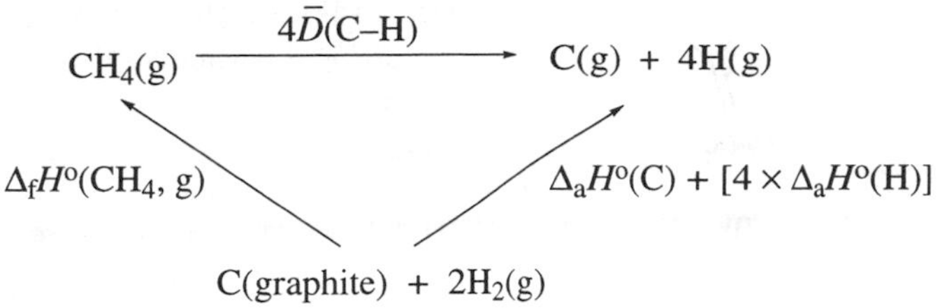

$$CH_4(g) \xrightarrow{\quad 4\bar{D}(C\text{–}H) \quad} C(g) + 4H(g)$$

$$\Delta_f H^\circ(CH_4, g) \qquad\qquad \Delta_a H^\circ(C) + [4 \times \Delta_a H^\circ(H)]$$

$$C(graphite) + 2H_2(g)$$

$$4\,\bar{D}(C\text{–}H) + \Delta_f H^\circ(CH_4, g) = \Delta_a H^\circ(C) + [4 \times \Delta_a H^\circ(H)]$$

$$4\,\bar{D}(C\text{–}H) = 717 + (4 \times 218) - (-74)$$

$$\bar{D} = \frac{1663}{4} = 415.75 \text{ kJ mol}^{-1}$$

This value can be transferred to ethene because $CH_4$ and $C_2H_4$ are similar compounds (they are both simple hydrocarbons). Set up a Hess cycle to find a value of $D(C=C)$ in ethene.

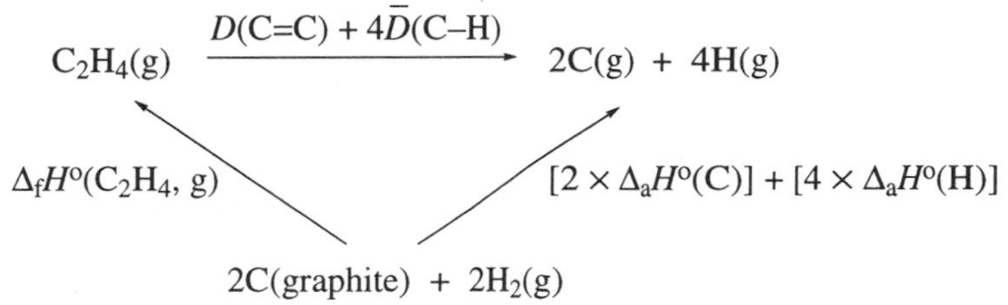

$$C_2H_4(g) \xrightarrow{\quad D(C=C) + 4\bar{D}(C\text{–}H) \quad} 2C(g) + 4H(g)$$

$$\Delta_f H^\circ(C_2H_4, g) \qquad\qquad [2 \times \Delta_a H^\circ(C)] + [4 \times \Delta_a H^\circ(H)]$$

$$2C(graphite) + 2H_2(g)$$

$$D(C=C) + 4\,\bar{D}(C\text{–}H) + \Delta_f H^\circ(C_2H_4, g) = [2 \times \Delta_a H^\circ(C)] + [4 \times \Delta_a H^\circ(H)]$$

$$D(C=C) + 4\,\bar{D}(C\text{–}H) = (2 \times 717) + (4 \times 218) - (52.5)$$

By substituting the value for the C–H bond dissociation enthalpy determined from data for methane, we have:

$$D(C=C) = (2 \times 717) + (4 \times 218) - (52.5) - (4 \times 415.75)$$

$$= 590.5 \text{ kJ mol}^{-1}$$

---

In the next problem set, we do not directly provide the data required but ask you to use the appendices in the main text *Chemistry: An Integrated Approach*. This exercise will make you think more carefully about the data needed for each calculation.

**Problem set 3.3**

1. Using appropriate data from Appendices 10 and 11 in H&C, estimate a value for $D(C\equiv C)$.

2. Using data from the Appendices of H&C, determine the average P–F bond dissociation enthalpy in (a) $PF_5$ and (b) $PF_3$. How do these values compare with each other? What conclusions can you draw?

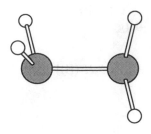

**3.3** The structure of $P_2H_4$.

3.  Estimate a value of $\overline{D}$(P–H) using data from the Appendices of H&C and, by transferring this value to $P_2H_4$ (Figure 3.3), estimate a value for $D$(P–P). How does this value compare with that obtained from $P_4$ in problem set 3.2, question 3?

4.  State, with reasons, whether you can reasonably transfer a value of the stated bond dissociation enthalpy between the following pairs of molecules: (a) the carbon-carbon bond enthalpy between ethene ($C_2H_4$) and ethyne ($C_2H_2$), (b) the sulfur–sulfur bond enthalpy between $S_8$, $S_{10}$ and $S_{18}$, (c) the sulfur-fluorine bond enthalpy between $SF_4$ and $SF_6$, and (d) the nitrogen-hydrogen bond enthalpy between ammonia ($NH_3$) and methylamine ($CH_3NH_2$). [*Hint*: Information about allotropes of sulfur is given in Table 3.2 in H&C.]

## LEWIS STRUCTURES

Although the main theme of this chapter is *homonuclear bonds*, we have introduced Lewis structures as one way of describing the bonding in homonuclear and heteronuclear molecules.

| Worked example 3.2 |

**Draw a Lewis structure for PCl₃.**

In what group of the periodic table is phosphorus? — Group 15
How many valence electrons does a P atom possess? — Five
[The ground state electronic configuration of P is [Ne]$3s^23p^3$].

In what group is chlorine? — Group 17
How many valence electrons does a Cl atom possess? — Seven
[The ground state electronic configuration of Cl is [Ne]$3s^23p^5$].

The Lewis structure of PCl₃ is

In this compound, each phosphorus and chlorine atom has an octet of valence electrons. [Lewis structures for $PF_3$, $PBr_3$ and $PI_3$ will be similar to that of $PCl_3$.]

| Problem set 3.4 |

1.  Draw Lewis structures for (a) $F_2$, (b) $O_2$, (c) $N_2$, (d) $H_2Se$, (e) $NF_3$, (f) $SnH_4$, (g) HBr, (h) $H_2O_2$ and (i) $C_2H_4$.

2.  Give Lewis representations of the structures of $CO_2$ and $SO_2$. In each compound, how many lone pairs does the central atom possess?

3.  Draw Lewis structures for (a) $BH_3$, (b) $NH_3$ and (c) $PH_3$. In which of these molecules does the central atom obey the octet rule?

4.  Draw Lewis structures for HCl, $Cl_2$, $ClF_3$ and $ClF_5$. In which molecules does chlorine obey the octet rule?

## VALENCE BOND AND MOLECULAR ORBITAL THEORIES

### Dihydrogen and derived ions

**VB = valence bond**
**MO = molecular orbital**

In problem set 3.5, we examine similarities and differences between the bonding in $H_2$ and the ions $[H_2]^+$ and $[H_2]^-$, and predict whether or not these ions may exist.

**Problem set 3.5**

1.  Draw a Lewis structure for $H_2$.
2.  By considering $[H_2]^-$ as a combination of H and $H^-$, draw a Lewis structure for $[H_2]^-$.
3.  How does VB theory describe the bonding in $H_2$?
4.  Construct an MO diagram to describe the bonding in $H_2$.
5.  Use the diagram from question 4 to determine (a) the bond order in $H_2$, and (b) whether $H_2$ is diamagnetic or paramagnetic.
6.  (a) Construct MO diagrams to describe the bonding in $[H_2]^+$ and $[H_2]^-$. (b) What is the bond order in each ion? (c) Would you expect these ions to exist? (d) Would the ions be diamagnetic or paramagnetic?
7.  Do $H_2$, $[H_2]^+$ and $[H_2]^-$ possess $\sigma$- or $\pi$-bonds?

### More diatomic species with $\sigma$-bonds

**Problem set 3.6**

1.  Draw a Lewis structure of $Li_2$ and use it to say what you can about the bonding in the molecule.
2.  Describe the bonding in $Li_2$ in terms of VB theory, including in your answer resonance structures, and an equation that describes the wavefunction $\psi_{molecule}$ for $Li_2$.
3.  Construct MO diagrams to describe the bonding in (a) $Li_2$, (b) $[Li_2]^+$, (c) $He_2$, (d) $Be_2$ and (e) $[He_2]^{2+}$. Write down the ground state electronic configuration of each species in a notational form, ignoring core electrons. For each species, what is the bond order? Which of these species do you predict might exist?
4.  The bond dissociation enthalpy of $Li_2$ is $110\ kJ\ mol^{-1}$. Do you expect the bond dissociation enthalpy of $[Li_2]^+$ to be greater or less than $110\ kJ\ mol^{-1}$? Rationalize your answer?
5.  The bond order in each of $Li_2$ and $F_2$ is one. Is it correct to assume that the bond dissociation enthalpies of these molecules will be similar?

**Problem set 3.7**

### $\sigma$- and $\pi$-orbitals

1.  Which of the atomic orbital combinations shown in Figure 3.4 are symmetry-allowed?
2.  Which of the symmetry-allowed combinations in Figure 3.4 lead to $\sigma$, $\sigma *$, $\pi$ or $\pi*$ interactions?
3.  What properties distinguish a $\sigma$- from a $\sigma *$-interaction?
4.  Consider two oxygen atoms defined to lie on the $z$ axis. Draw a diagram to represent the molecular orbital resulting from the in-phase overlap between (a) two $2p_z$ atomic orbitals, and (b) two $2p_x$ atomic orbitals.

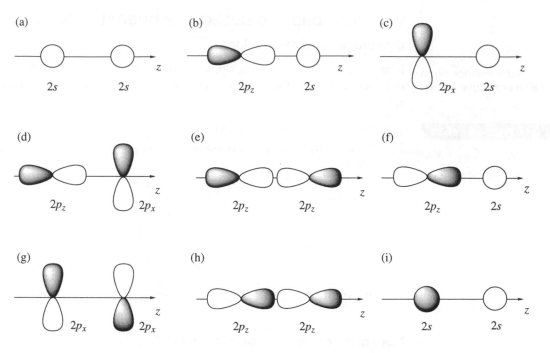

**3.4**  Orbital interactions for questions 1 and 2 in problem set 3.7.

## Diatomics with $\sigma$- and $\pi$-bonds

In the main text, we looked at the bonding in the diatomic molecules $B_2$, $C_2$, $N_2$ and $O_2$ to illustrate $\sigma$- and $\pi$-orbital interactions. We showed that the relative ordering of the orbitals in the MO diagram may change because of the effects of orbital mixing.

The following questions test your general understanding of the MO treatment of the bonding in these and similar species.

**Problem set 3.8**

1.  What do you understand by the term *linear combination of atomic orbitals* (LCAO)?

2.  Draw a *general* MO diagram (without electrons) which shows the interactions of the valence $2s$ and $2p$ AOs of two X atoms (defined to lie on the $z$ axis) assuming that a simple LCAO method is appropriate.

3.  How does the MO diagram you have drawn in question 2 change if you allow for mixing between the $\sigma(2s)$ and $\sigma(2p)$ and between the $\sigma^*(2s)$ and $\sigma^*(2p)$ MOs? Draw a revised MO diagram for the molecule $X_2$.

4.  Use the MO diagram drawn in question 2 to deduce (a) the bond order, and (b) the magnetic properties of the molecule $B_2$. Repeat the exercise using the MO diagram you have drawn in question 3. What *experimental* evidence is available to distinguish between these MO diagrams as an appropriate

way in which to describe the bonding in $B_2$?

5. Would you expect orbital mixing to be important in $F_2$? Explain your answer.

6. A description of the bonding in a diatomic molecule $X_2$ can be approached by drawing a Lewis structure, using valence bond theory, or constructing an MO diagram. Assess the relative merits of these three methods for the specific case of $O_2$.

7. Figure 3.5 shows the trends in bond dissociation enthalpies along the series (a) $O_2$, $[O_2]^-$ and $[O_2]^{2-}$, (b) $Li_2$, $Na_2$ and $K_2$, and (c) $F_2$, $Cl_2$, $Br_2$, $I_2$. Using whatever bonding models seem most appropriate, discuss these trends, paying particular attention to the *different* factors that may contribute to a decrease in bond dissociation enthalpy.

**3.5** Data for question 7, problem set 3.8.

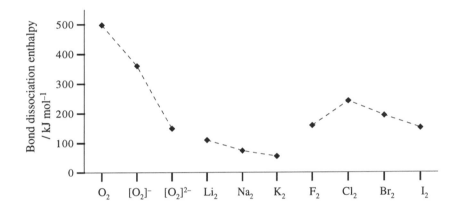

## Advanced problems

Answers for the following problems are not provided.

**Problem set 3.9**

1. Explain why oxygen and nitrogen form dinuclear molecules whereas sulfur and phosphorus exist as polynuclear species.

2. Would you expect the structures of the heavier congeners (i.e. lower members) of groups 15 and 16 to resemble nitrogen and oxygen, or phosphorus and sulfur respectively?

3. Considering your answers to questions 1 and 2 above, why do all of the elements of group 17 exist as dinuclear species?

4. Will all of the elements of group 18 always exist as mononuclear species?

5. (a) Construct a molecular orbital diagram for $Ne_2$. Would you expect $Ne_2$ molecules to exist? (b) In reality, an $Ne_2$ molecule would be *less stable* than two Ne atoms. Does your MO diagram indicate this, and if not, how might you modify it to account for this fact. (c) What are the physical significances of any changes that you have made to the diagram in answer to part (b)?

6.    (a) Construct an MO diagram for the formation of $O_2$ from two O atoms.
      (b)  Now try to construct a similar MO diagram for $O_3$. What problems do you encounter? [We discuss MO diagrams for polyatomic molecules in Chapter 5.]

7.    (a) Experimentally $O_2$ is found to be paramagnetic. Account for this fact.
      (b) Two different excited states of $O_2$ are known which are diamagnetic. Use an MO diagram to show possible electronic configurations that would account for these observations.

# 4 Heteronuclear diatomic molecules

**Topics**
- Bonding in molecules with $\sigma$-bonds
- Electronegativity
- Dipole moments
- Isoelectronic species
- Bonding in molecules with $\sigma$- and $\pi$-bonds

## BONDING IN MOLECULES WITH $\sigma$-BONDS

### Valence bond theory

In the next set of questions, we begin by considering *homonuclear* diatomic molecules before moving to heteronuclear diatomics.

**Problem set 4.1**

1. Draw resonance structures to represent the bonding in $F_2$. Which resonance structure(s) predominate(s)?

2. The equation:

$$\psi_{molecule} = \left\{ c_1 \times \psi_{covalent} \right\} + \left\{ c_2 \times \psi_{ionic} \right\}$$

describes the bonding in $Cl_2$ using VB theory. Which of the following statements is correct? (a) $c_1 = c_2$; (b) $c_1 < c_2$; (c) $c_1 > c_2$.

3. Write down resonance structures to describe the bonding in the molecules (a) HF, (b) NaH, and (c) ClF. Suggest which structures contribute the most to the overall bonding picture for each molecule.

### The hydroxide ion [OH]$^-$

In the main text, we considered the bonding in HF in terms of a Lewis structure, VB theory and the MO approach. The following set of problems guides you through an analysis of the bonding in the hydroxide ion.

**Problem set 4.2**

1. Write down the ground state electronic configuration of the O$^-$ ion. How is this related to that of the fluorine atom? How is the [OH]$^-$ ion related to the HF molecule?

2. Draw a Lewis structure for the [OH]$^-$ ion.

3. Using the valence bond model, describe the bonding in the [OH]$^-$ ion in terms of a set of resonance structures. Are all the resonance structures equally important?

4. What does the Lewis structure tell you about (a) the bond order, and (b) the

magnetic properties of the hydroxide ion?

5.   Construct an approximate MO diagram for the bonding in [OH]⁻ by considering the interaction of the valence atomic orbitals of the O⁻ ion and an H atom. Before you draw the diagram, look back at your answer to question 1, and think about the effect that the negative charge has on the energies of the AOs of the O⁻ ion.

6.   Using the MO diagram that you have drawn for question 5, determine (a) the bond order, (b) the magnetic properties and (c) the character of the σ-bonding MO in [OH]⁻.

7.   Compare the results of the MO analysis of the bonding in [OH]⁻ with the MO description of HF. How do you account for any similarities and differences between the bonding in these species?

➤
**MO picture for HF: see Section 4.5 in H&C**

## Bonding in HCl

**Problem set 4.3**

➤
**The electron volt (eV) is a unit of energy.**
**1 eV = 96.5 kJ mol⁻¹**

1.   The results of a theoretical study of the HCl molecule show that the energies of the valence Cl $3s$ and $3p$ AOs are $-29$ and $-13$ eV respectively, and the H $1s$ AO lies at $-14$ eV. Using these data and an LCAO approach, construct an approximate MO diagram to show the orbital interactions in HCl.

2.   Use the diagram to determine (a) the bond order in HCl, (b) whether HCl is diamagnetic or paramagnetic, (c) the character of the $\sigma$ MO, and (d) the character of the $\sigma^*$ MO. Sketch representations of these latter MOs.

3.   *Advanced study question.* Check the answer to question 1; the answer given assumes the simplest of pictures, and only allows for the overlap of the H $1s$ with the Cl $3p$ since these are closer in energy than the H $1s$ and Cl $3s$. Figure 4.1 shows an MO diagram for HCl which is the result of a computational study. Account for the differences between this diagram and the one given as the answer for question 1. [*Hint*: The discussion of CO in H&C, Section 4.14 is relevant to this question.]

**4.1** MO diagram for the formation of HCl (for use in question 3, problem set 4.3).

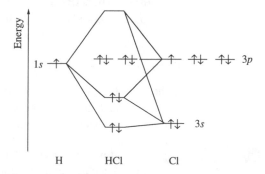

## ELECTRONEGATIVITY

Electronegativity, $\chi^P$, was defined by Pauling as 'the power of an atom in a molecule to attract electrons to itself'.

In a bond X—Y, if atom Y is more electronegative than X, the electrons in the bonding region are attracted towards atom Y, and the resonance structure X⁺Y⁻

makes a significant contribution to the overall bonding picture. Values of Pauling electronegativities are given in Appendix 7 of H&C.

Pauling electronegativity values are estimated from bond dissociation enthalpy data, and conversely, the dissociation enthalpy of a heteronuclear bond can be estimated by using equations 4.1 and 4.2.

➤
**These equations are obtained by rearranging equations 4.15 and 4.16 in Section 4.6 of H&C**

$$D(X-Y) = \left\{ \frac{1}{2} \times [D(X-X) + D(Y-Y)] \right\} + \Delta D \qquad (4.1)$$

where:

$$\Delta D = \left\{ \chi^P(Y) - \chi^P(X) \right\}^2 \qquad (4.2)$$

The units of $\Delta D$ in equation 4.2 are electron volts (eV); 1 eV = 96.5 kJ mol$^{-1}$.

| Worked example 4.1 |

**Estimate the dissociation enthalpy of the bond in hydrogen bromide using the following data.**
**$D$(H–H) = 436 kJ mol$^{-1}$         $D$(Br–Br) = 193 kJ mol$^{-1}$**
**$\chi^P$(H) = 2.2                        $\chi^P$(Br) = 3.0**

The equations needed are 4.1 and 4.2 (above). First, determine the value of $\Delta D$:

$$\Delta D = \left\{ \chi^P(Br) - \chi^P(H) \right\}^2$$

$$\Delta D = \{3.0 - 2.2\}^2 = 0.64 \text{ eV}$$

Now convert this enthalpy value into kJ mol$^{-1}$:

$$1 \text{ eV} = 96.5 \text{ kJ mol}^{-1}$$
$$0.64 \text{ eV} = 0.64 \times 96.5 = 61.76 \text{ kJ mol}^{-1}$$

Now, find $D$(H–Br):

$$D(H-Br) = \left\{ \frac{1}{2} \times [D(H-H) + D(Br-Br)] \right\} + \Delta D$$

$$= \left\{ \frac{1}{2} \times [436 + 193] \right\} + 61.76$$

$$= 376.26 \approx 376 \text{ kJ mol}^{-1}$$

[This compares with an experimentally determined value of 366 kJ mol$^{-1}$.]

| Problem set 4.4 |
➤
**Values of $\chi^P$:
see Appendix 7 in H&C**

1.  Given that the bond dissociation enthalpies of $F_2$ and $Cl_2$ are 159 and 242 kJ mol$^{-1}$ respectively, estimate a value for $D$(Cl–F) in chlorine monofluoride.

2.  Why is it not reasonable to assume a simple additivity rule in question 1 and estimate $D$(Cl–F) using the following equation?

$$D(Cl-F) = \left\{ \frac{1}{2} \times [D(Cl-Cl) + D(F-F)] \right\}$$

3.   It is possible to apply equations 4.1 and 4.2 to heteronuclear bonds in polyatomic molecules. Given that $D(\text{O–O})$ and $D(\text{H–H})$ are 146 and 436 kJ mol$^{-1}$ respectively, estimate the bond dissociation enthalpy of an O–H bond.

4.   Set up a suitable Hess cycle to determine $D(\text{O–H})$ in a water molecule, given that $\Delta_f H°(\text{H}_2\text{O, g}) = -242$ kJ mol$^{-1}$; further data required are listed in Appendix 10 of H&C. Why is it more appropriate to use *gaseous* rather than liquid $\text{H}_2\text{O}$ in this thermochemical cycle? Compare the value of $D(\text{O–H})$ obtained by this method with that obtained in question 3.

5.   Why are Pauling electronegativity values dependent on oxidation state, e.g. $\chi^P$ for Pb(II) is 1.9 and for Pb(IV) is 2.3?

6.   Rationalize the trends in electronegativity values shown in Figures 4.2 and 4.3.

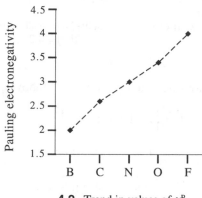

**4.2**  Trend in values of $\chi^P$ for elements from $Z = 5$ to 9.

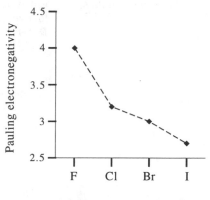

**4.3**  Trend in values of $\chi^P$ for elements in group 17.

## ELECTRIC DIPOLE MOMENTS

In this section, the use of 'dipole moment' refers to an *electric* dipole moment, the symbol for which is $\mu$.

For the following problems, you will need to refer to Appendix 7 of H&C for the table of Pauling electronegativity values .

**The unit of $\mu$ is coulomb metre (C m) or debye (D). 1 D = $3.336 \times 10^{-30}$ C m**

**Problem set 4.5**

1.   (a) What is the difference between a vector and a scalar quantity? (b) Is a dipole moment a vector or a scalar quantity?

2.   How does a dipole moment in a heteronuclear diatomic molecule arise?

3.   Which of the following molecules is polar (a) $\text{Cl}_2$; (b) HCl; (c) ClF; (d) $\text{H}_2$; (e) HI? Draw a diagram of each polar molecule showing in which direction the dipole moment acts.

4.   In each of the following gas phase molecules, which atom carries the $\delta^+$ charge (a) HBr; (b) IBr; (c) BrF; (d) IF; (e) RbCl; (f) TlF?

## ISOELECTRONIC SPECIES

The term *isoelectronic* means the 'same number of electrons'. Strictly it refers to the *total* number (core + valence) of electrons. However, we often refer to molecules (or ions) as being isoelectronic when they possess the same number of valence electrons but different numbers of core electrons. In these cases, we should say 'isoelectronic with respect to the valence electrons'.

The concept of isoelectronic species is a very useful one, allowing you to spot species that *may* be similar in respect of their chemical properties. For example, the halogen atoms all belong to the same periodic group, all possess seven valence electrons, and are therefore isoelectronic (with respect to their valence electrons); chemically, the halogens tend to behave in similar ways. However, we must be careful! On descending group 17, higher oxidation states become increasingly stable, e.g. iodine forms fluorides up to $IF_7$, chlorine forms fluorides up to $ClF_5$, but fluorine only forms $F_2$.

> **Chemistry of the halogens: see Sections 11.11 and 13.12 in H&C**

**Problem set 4.6**

1.  Consider the elements in the periodic rows from boron to neon, and aluminium to argon. Write down the ground state electronic configuration for each atom. (a) How many electrons would a nitrogen atom have to gain in order to become isoelectronic with a neon atom? (b) With which atom is an $S^{2-}$ ion isoelectronic? (c) With which atom is an $Al^{3+}$ isoelectronic? (d) Write down three atoms or ions that are isoelectronic with an $O^{2-}$ ion. (e) Is $Se^{2-}$ isoelectronic with $O^{2-}$? (f) What charge must a phosphorus centre carry in order for the species to be isoelectronic with a $Cl^-$ ion?

2.  With which singly charged cation is the $Ca^{2+}$ ion isoelectronic?

3.  With which doubly charged anion is the $Na^+$ ion isoelectronic?

4.  With which noble gas is the $Sr^{2+}$ ion isoelectronic?

5.  (a) How many *core* electrons does the molecule $Cl_2$ possess? (b) How many *valence* electrons does $Cl_2$ possess? (c) Which of the following species are isoelectronic with $Cl_2$: $[O_2]^{2-}$, $[S_2]^{2-}$, $Br_2$, $S_2$? Qualify your answer to part (c) stating how you are using the term 'isoelectronic'.

6.  Rationalize why CO, $[CN]^-$ and $N_2$ are isoelectronic species.

7.  We can extend the isoelectronic principle to polyatomic species. Which of the following species are isoelectronic with $CH_4$: (a) $[BH_4]^-$, (b) $CCl_4$, (c) $[NH_4]^+$, (d) $SiH_4$, (e) $[AlH_4]^-$, (f) $CH_3F$. Qualify your answer stating in what context you are using the term 'isoelectronic'.

## BONDING IN MOLECULES WITH $\sigma$- AND $\pi$-BONDS

You have probably noticed that in our discussions of heteronuclear diatomic molecules with $\sigma$-bonds we only considered a small range of molecules, for example, the hydrogen halides (HF, HCl, HBr, HI) and some interhalogen compounds (e.g. ClF, BrF, IBr). The reason should be clear — each of the halogen atoms needs one more electron to complete its octet and so forms a single bond either with hydrogen or another halogen atom. In this section, we extend the

> **Interhalogen compounds: see Section 13.12 in H&C**

discussion to heteronuclear diatomic species containing multiple bonds. In the main text, we focussed on the bonding in carbon monoxide. Here we consider nitrogen monoxide (NO) and chlorine monoxide (ClO). The discussion is in the form of two 'hands-on' exercises, and as you work through the following problem sets you will build up information about the properties of these diatomic molecules. Compare the results for NO with those obtained in Section 4.16 of H&C where we used the MO diagram for CO to gain insight into the bonding in a series of related species.

## NO: A molecule of biological importance but a contributor to atmospheric pollution

**Problem set 4.7**

➤

**NO plays an important role in biology: see Box 4.5 in H&C**

**NO$_x$ is a pollutant: see Section 13.8 in H&C and page 118 of the workbook**

1.  Draw a Lewis structure for NO.
2.  Draw resonance structures for NO that you consider will contribute significantly to the bonding in NO.
3.  What conclusions can you draw from VB theory about (a) the bond order, and (b) the magnetic properties of NO?
4.  Purely on the basis of Pauling electronegativity values, do you think that NO is a polar molecule? Which atom would you predict would carry the $\delta^-$ charge? In view of the resonance structures that you have drawn in question 2, are you confident about your decision concerning the charge distribution in NO? [*Hint*: Look at the discussion about CO in Section 4.13 of H&C.]
5.  Write down the ground state electronic configurations of the N and O atoms. What do you think will be the *relative* energies of the valence orbitals of the two atoms? [Think about the factors influencing orbital energy.]
6.  Using the result of question 5, construct an approximate MO diagram for the formation of NO assuming a simple LCAO approach.
7.  Using the diagram from question 6, deduce (a) whether NO is diamagnetic or paramagnetic, (b) the bond order in NO, (c) the character of the lowest energy molecular orbital, (d) the character of the highest occupied molecular orbital. Illustrate your answers to (c) and (d) by sketching diagrams of the orbitals.
8.  *Advanced study question.* A theoretical study of NO gives the following orbital energies for N and O: N(2s) –27 eV, N(2p) –13 eV, O(2s) –35 eV, O(2p) –17 eV. The relative ordering of these orbitals results in a significant amount of mixing of orbital character in the NO molecule. Figure 4.4 shows an MO diagram that has been constructed on the basis of the theoretical study; the diagram is over-simplified. What differences do you notice between your MO diagram (from question 6) and that in Figure 4.4? How do you account for these differences? What features about the bonding description of NO remain *unchanged* when you compare the two diagrams? How valid was your simple approach?
9.  (a) What is the product when NO undergoes a one-electron oxidation? (b) Write an equation to represent this process. (c) How can you represent the oxidation in terms of Figure 4.4?

**4.4** An MO diagram for the formation of nitrogen monoxide. The diagram shows that some orbital mixing occurs, although it is over-simplified.

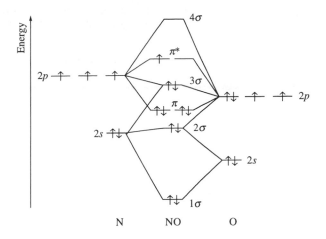

## CIO: A product of ozone depletion

➤

**Depletion of the ozone layer: see Box 8.9 in H&C**

Chlorine radicals produced in the stratosphere by the degradation of chlorofluorocarbons (CFCs) react with ozone to give the radical chlorine monoxide:

$$O_3 + Cl^• \rightarrow O_2 + ClO^•$$

In the questions below, we write ClO without explicitly showing that it is a radical.

**Problem set 4.8**

1. Draw a Lewis structure for ClO. Does this model suggest that the odd electron resides on the chlorine or oxygen atom?
2. ClO undergoes dimerization to $Cl_2O_2$. Suggest a structure for the dimer.
3. When the concentration of ClO in the stratosphere is high, the dimer $Cl_2O_2$ is likely to form, but this decomposes in sunlight:

$$Cl_2O_2 \rightarrow ClOO + Cl$$

Further decomposition occurs to give dioxygen:

$$ClOO \rightarrow Cl + O_2$$

Draw a Lewis structure for ClOO.
4. From VB theory, what is the bond order in ClO?
5. Write down the ground state electronic configurations of the Cl and O atoms.
6. Although the valence orbitals of the Cl atom are the $3s$ and $3p$ and those of the O atom are the $2s$ and $2p$, their relative orbital energies can be represented as in Figure 4.5. Using the information in Figure 4.5, construct an approximate MO diagram for the formation of ClO.

7.    Using the diagram from question 6, deduce the bond order in ClO. Does this result agree with that from VB theory?

8.    Using the diagram from question 6, deduce the approximate character of (a) the $\sigma(s)$ MO, and (b) the orbital containing the unpaired electron. Does MO theory suggest that the odd electron is localized on either of the Cl or O atoms?

**4.5** The approximate relative energies of the valence AOs of chlorine and oxygen before combination to give ClO.

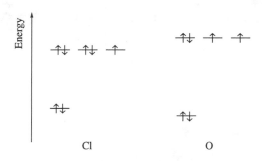

# 5 Polyatomic molecules: shapes and bonding

---

**Topics**
- The valence-shell electron-pair repulsion (VSEPR) model
- The Kepert model
- Geometrical isomerism
- Five-coordinate structures
- Molecular dipole moments
- Expansion of the octet and resonance structures
- Hybridization of atomic orbitals

---

## THE VALENCE-SHELL ELECTRON-PAIR REPULSION (VSEPR) MODEL

The VSEPR model considers the repulsions between electron pairs in the valence shell of the central atom of a molecule or ion, and the mutual repulsions follow the sequence:

$$\left(\begin{array}{c}\text{Lone pair–lone pair}\\ \text{repulsions}\end{array}\right) > \left(\begin{array}{c}\text{Lone pair–bonding pair}\\ \text{repulsions}\end{array}\right) > \left(\begin{array}{c}\text{Bonding pair–bonding pair}\\ \text{repulsions}\end{array}\right)$$

The shapes to which we refer in this section are illustrated in Figures 5.9 and 5.12 of H&C. You should ensure that you are familiar with the terms:
- linear
- trigonal planar
- tetrahedral
- trigonal bipyramidal
- octahedral

and that you can draw the shapes of molecules of general formula $XY_2$, $XY_3$, $XY_4$, $XY_5$ and $XY_6$ that possess these geometries.

**Worked example 5.1** | **Predict the shape of a molecule of phosphorus(III) chloride, $PCl_3$.**

Phosphorus is in group 15 and has five valence electrons.
Chlorine is in group 17 and has seven valence electrons.
The Lewis structure for the $PCl_3$ molecule is:

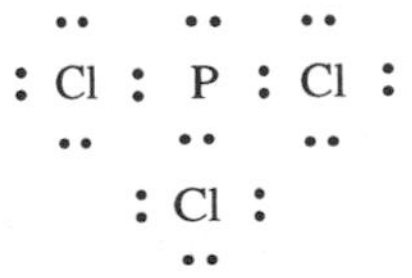

$$: \overset{\cdot\cdot}{Cl} : \overset{\cdot\cdot}{P} : \overset{\cdot\cdot}{Cl} :$$
$$: \overset{\cdot\cdot}{Cl} :$$

There are four points of negative charge around the phosphorus atom (three bonding pairs and one lone pair of electrons).

➤ Angles in tetrahedral molecules: see H&C Figure 5.18

The geometry of the $PCl_3$ molecule is derived from a tetrahedron; the four atoms define a trigonal pyramidal structure (Figure 5.1). A value of ∠Cl-P-Cl < 109.5° is predicted. (The experimentally determined Cl-P-Cl bond angle is 100.3°). [*Question*: Do the lone pairs of electrons on the chlorine atoms influence the shape of the $PCl_3$ molecule?].

**5.1** The predicted structure of $PCl_3$. *With* the lone pair, it may be described as being based on a tetrahedron, but the *molecular structure* is trigonal pyramidal.

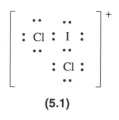

---

| Worked example 5.2 | **Predict the structure of the cation [ICl$_2$]$^+$.** |

Iodine is in group 17 and has seven valence electrons. The positive charge may be assigned to the iodine atom; $I^+$ has six valence electrons.

Chlorine is in group 17 and has seven valence electrons.

Draw the Lewis structure for the [ICl$_2$]$^+$ ion — structure **5.1**.

There are four points of negative charge around the iodine atom (two bonding pairs of electrons and two lone pairs of electrons).

The geometry of the [ICl$_2$]$^+$ ion is derived from a tetrahedron and the three atoms define a bent shape (Figure 5.2). The Cl-I-Cl bond angle is predicted to be less than 109.5°; the experimental value is 91.5°.

**(5.1)**

**5.2** The predicted structure of the cation [ICl$_2$]$^+$. *With* the lone pairs, it is described as being based on a tetrahedron; the cation itself is bent.

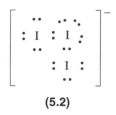

---

| Worked example 5.3 | **Predict the shape of the [I$_3$]$^-$ anion.** |

Iodine is in group 17 and has seven valence electrons. The negative charge may be assigned to the central iodine atom; $I^-$ has eight valence electrons.

Draw the Lewis structure for the [I$_3$]$^-$ ion — structure **5.2**.

There are five points of negative charge around the central iodine atom (two bonding and three lone pairs of electrons).

The geometry of the [I$_3$]$^-$ ion is derived from a trigonal bipyramid. The three lone pairs prefer be as far apart as possible. Placing them in the equatorial plane minimizes repulsions; all lone pair-lone pair angles are 120° (Figure 5.3).

The [I$_3$]$^-$ ion is linear as shown in Figure 5.3.

**(5.2)**

**5.3** The predicted structure of the anion [I$_3$]$^-$. *With* the lone pairs, it is described as being based on a trigonal bipyramid, but the anion itself is linear.

1.  Using VSEPR theory, predict the structures of the molecules (a) $SiCl_4$; (b) $BF_3$; (c) $SF_6$; (d) $AsF_5$; (e) $IF_5$; (f) $OF_2$; (g) $SOF_4$; (h) $POCl_3$.
2.  Using the VSEPR model, predict the structures of the following ions (a) $[TeF_5]^-$; (b) $[ICl_2]^-$; (c) $[BF_4]^-$; (d) $[PF_4]^+$; (e) $[SbF_6]^-$; (f) $[ClO_4]^-$; (g) $[PO_4]^{3-}$; (h) $[H_3O]^+$.
3.  Suggest approximate values (or limiting values) for the stated bond angles in the following species: (a) the H–O–H bond angles in $[H_3O]^+$; (b) the H–S–H bond angles in $H_2S$; (c) the F–Kr–F bond angles in $KrF_2$; (d) the Cl–I–Cl bond angles in $[ICl_4]^-$; (e) the C–N–C bond angles in $[NMe_4]^+$ (Me = $CH_3$); (f) the O–S–O bond angles in $[SO_4]^{2-}$.

## THE KEPERT MODEL

The Kepert model is concerned with the mutual repulsions between groups attached to a *d*-block metal centre. *Steric effects* are taken into consideration. Unlike the VSEPR model, lone pairs of electrons on the central atom are ignored. As well as looking at simple $ML_n$ species such as $[Au(CN)_2]^-$ or $[FeF_6]^{3-}$ where the ligands are identical, the Kepert model can be used to suggest which sites might be occupied by ligands with differing *steric demands*, for example in $[NiBr_3(PMe_2Ph)_2]$, Figure 5.20 in H&C.

**Worked example 5.4**

**Predict the structure of the cation $[Au\{PPh(C_6H_{11})_2\}_3]^+$.**

➤
**Ph = phenyl = $C_6H_5$**
**$C_6H_{11}$ = cyclohexyl**

$$\left[ L-Au\underset{L}{\overset{\cdots\text{\tiny III} L}{\blacktriangleleft}} \right]^+$$

**(5.3)**

Despite the $PPh(C_6H_{11})_2$ groups appearing to be rather complicated, this cation is of the simple form $[AuL_3]^+$.

By the Kepert model, three groups will adopt a trigonal planar arrangement around the metal centre and structure **5.3** shows the predicted structure of a cation of type $[AuL_3]^+$. (Compare this with the experimental results, detailed below.)

The structure of the salt $[Au\{PPh(C_6H_{11})_2\}_3][ClO_4]$ has been determined by X-ray diffraction methods and the cation is shown in Figure 5.4. The gold centre is in the predicted trigonal planar environment, but the figure also shows how the particular arrangement of the ligands further minimizes inter-ligand repulsions; the phenyl groups are positioned in between the more sterically demanding cyclohexyl groups. Notice the 'propeller-like' arrangement of the ligands which helps to minimize inter-ligand repulsions.

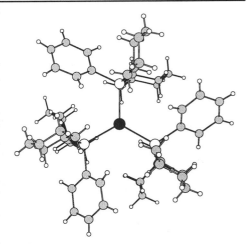

**5.4** The structure of the cation $[Au\{PPh(C_6H_{11})_2\}_3]^+$ determined by X-ray diffraction methods.

**Problem set 5.2**

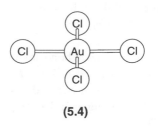

**(5.4)**

1.  Using the Kepert model, predict the shapes of the following species involving a central *d*-block metal: (a) $[FeO_4]^{2-}$; (b) $[W(CO)_6]$; (c) $[Cr(NH_3)_6]^{3+}$; (d) $[Ag(NH_3)_2]^+$; (e) $[Fe(CN)_6]^{3-}$; (f) $[Mn(CO)_5]^-$; (g) $[ZnCl_4]^{2-}$; (h) $[HgCl_2]$; (i) $[HgI_3]^-$; (j) $[MnO_4]^-$.

2.  The shape of the $[AuCl_4]^-$ anion is shown in structure **5.4**. Is this as predicted by the Kepert model?

3.  For eight-coordinate structures, three possible polyhedra are shown in Figure 5.5. Of these, the cube and the square-antiprism are related to each other. (a) How would you describe this relationship? (b) Using Kepert's approach, suggest whether $[Mo(CN)_8]^{4-}$ is likely to adopt a cubic or square-antiprismatic arrangement of cyano-groups.

**5.5** Three possible arrangements of the eight L groups in $ML_8$ species.

Cube          Square-antiprism          Dodecahedron

## A MISCELLANY OF STRUCTURES

In the following problem set, we combine the ideas of the VSEPR and Kepert models, and ask you to choose the appropriate model for a given situation.

**Problem set 5.3**

1.  Rationalize why $CO_2$ is linear but $SO_2$ is bent.
2.  Rationalize why the ions $[Br_3]^-$ and $[AuCl_2]^-$ are linear.
3.  Rationalize why $SnCl_4$ is tetrahedral but $[ICl_4]^-$ is square planar.
4.  For each of the structures shown in Figure 5.6, rationalize the arrangement of the atoms or groups around the central atom.

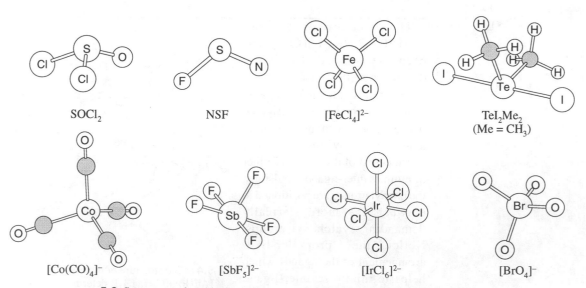

**5.6** Structures of molecules and ions needed for question 4, problem set 5.3.

## GEOMETRICAL ISOMERISM

### Molecules and anions with single atomic centres

1. What are *geometrical isomers*?
2. What is the molecular shape of each of the following, and which molecules possess geometrical isomers: (a) $BBrCl_2$; (b) $PBrCl_2$; (c) $PCl_3F_2$; (d) $PClF_4$; (e) $SClF_5$?
3. How many geometrical isomers does $WCl_2F_4$ possess? Draw their structures and assign appropriate labels (prefixes) to them.
4. How many geometrical isomers does $WCl_3F_3$ possess? Draw their structures and assign appropriate labels to them.
5. How many geometrical isomers does the cation $[Ru(NH_3)_4Br_2]^+$ possess? [Each $NH_3$ molecule bonds to the metal centre through the nitrogen atom.]
6. In $[Pt(NH_3)_2Cl_2]$, the platinum centre is in a square planar environment. Draw the *cis* and *trans* isomers of this compound.
7. Although we have looked at *trans*- and *cis*-square planar complexes in terms of the general formulae $MX_2Y_2$, a square planar complex $MX_2YZ$ also possesses *trans*- and *cis*-isomers. Draw the structure of Vaska's compound, *trans*-$[Ir(PPh_3)_2(CO)Cl]$.

### Compounds with double bonds

Compounds containing a double bond *may* possess geometrical isomers.

1. Which of the following molecules contain double bonds? (a) $H_2O_2$; (b) $N_2F_2$; (c) $N_2F_4$; (d) $C_2H_4$; (e) ClNO; (f) $COCl_2$. [*Hint*: Draw Lewis structures of the molecules to help you.]
2. Of those molecules in question 1 that *do* possess a double bond, which possess geometrical isomers?
3. To what do the prefixes (*E*) and (*Z*) refer?
4. Dichloroethene has the molecular formula $C_2H_2Cl_2$. How many isomers of this compound are there? Which of the isomers are *geometrical* isomers of each other?
5. A representation of the structure of the dye Sudan red B is shown in Figure 5.7. What is wrong with this structural representation? Redraw the structure so as to give a better picture of the molecule. [This is not a simple matter!]

**5.7** A representation of the structure of the dye Sudan red B.

➤
**Azo-dyes:**
**see H&C, Section 9.11**

## FIVE-COORDINATE STRUCTURES

➤
**NMR = nuclear magnetic resonance**

The behaviour of *stereochemically non-rigid* species in solution is often observed by using NMR spectroscopy. Details of this technique are described in Chapter 9, but here we look at the results of experiments that illustrate such behaviour.

**Problem set 5.6**

1. How many different fluorine environments are there in a static molecule of $PF_5$? By what names are the sites usually differentiated?
2. Figure 5.8 shows two representations of a $PF_5$ molecule and the F atoms are numbered 1 to 5. How is it possible to convert the left-hand structure into the right-hand one? During the interconversion, are any of the P–F bonds broken?
3. What is the name given to the process you have described in question 2? Give another example of a molecule that undergoes this dynamic process in solution.
4. The results of a $^{19}F$ NMR spectroscopic investigation of $PF_5$ at room temperature indicate that the $PF_5$ molecule only possesses one fluorine environment. By referring to Figure 5.8, explain how *all* the fluorine atoms are apparently equivalent.
   [We re-examine the NMR spectra of $PF_5$ on page 88 of the workbook.]

**5.8** Interconversion of atom sites in $PF_5$, an example of a five-coordinate molecule.

## MOLECULAR DIPOLE MOMENTS

*Remember* ➤ A dipole moment is a *vector quantity* — it has *magnitude and direction*.

**Worked example 5.5**

**Do you expect $SF_4$ to be a polar or non-polar molecule?**

First, use VSEPR theory to draw the structure of $SF_4$. The sulfur atom has six valence electrons and, after forming four S–F single bonds, the sulfur has a lone pair of electrons remaining:

➤
**For a detailed description of the structure, see example 5 in Section 5.6 in H&C**

Now consider the electronegativities of S and F; Appendix 7 in H&C gives $\chi^P(S) = 2.6$, $\chi^P(F) = 4.0$. Each S–F *bond* is therefore polar:

$$S—F$$

Now consider the $SF_4$ molecule as a whole.

*Axial S–F bond dipole moments*:   These two bond dipoles cancel each other because they are of equal magnitude but act in opposite directions.

*Equatorial S–F bond dipole moments*:   These reinforce each other and their resultant acts in a direction opposite to the lone pair:

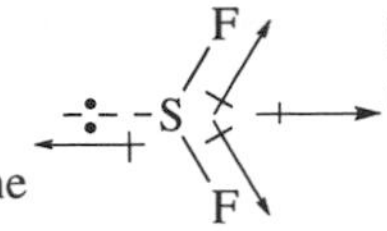

You now have to assess whether there will be a net dipole moment and this depends upon the relative magnitudes of the dipole moments shown above. Without more information, this is not easy to do. [The experimental value of $\mu$ for $SF_4$ is 0.63 D.]

---

**Problem set 5.7**

**Refer to Pauling electronegativity values in Appendix 7 of H&C**

1.   Which of the following molecules do you expect to be polar and why? (a) HCN; (b) $AsF_3$; (c) $AlCl_3$; (d) $CHBr_3$; (e) $CS_2$; (f) $SO_2$; (g) OCSe; (h) $P_4$.

2.   The dipole moment of (Z)-$N_2F_4$ in the gas phase is 0.16 D, but (E)-$N_2F_4$ is non-polar. Explain this difference.

3.   The dipole moments of the gas phase molecules $H_2O$ and $F_2O$ are 1.85 and 0.30 D. Rationalize the relative magnitudes of these values.

4.   For the gas phase molecule $CHBr_3$, the measured dipole moment is 0.99 D. Draw the structure of $CHBr_3$ and show the direction in which the dipole moment acts.

5.   The two allotropes of oxygen are $O_2$ and $O_3$. One is non-polar and one has a dipole moment of 0.53 D. Assign these polarity data and explain this difference in properties.

---

## EXPANSION OF THE OCTET AND RESONANCE STRUCTURES

**Problem set 5.8**

1.   In which of the following molecules does the central atom obey the octet rule (a) $BCl_3$; (b) $NF_3$; (c) $PCl_3$; (d) $PF_5$; (e) $SF_4$; (f) $SF_6$; (g) $OF_2$; (h) ClF; (i) $ClF_3$; (j) $ClF_5$; (k) $XeF_2$?

2.   In which of the molecules in question 1 is there a *sextet* electrons around the central atom?

3.   In *general*, which elements tend to obey the octet rule when they form covalent bonds? *Why* is an octet the 'magic number' for these elements?

4.   For those elements which are able to expand their octet of valence electrons, what is the *maximum* number of electrons that may be accommodated in the valence shell and why? Is this number of electrons often attained?

5.   Nitrogen tends to obey the octet rule. How many single bonds may an N atomic centre form? How do you explain the fact that in the $[NH_4]^+$ ion this number is exceeded?

6.   How many *single bonds* may the following centres form whilst obeying the octet rule (a) $O^-$; (b) $O^+$; (c) $B^-$; (d) $N^+$; (e) $N^-$.

When you draw resonance structures, you must
- keep in mind the bonding capabilities of each atom; for example, a neutral O centre may only form two single bonds or one double bond, and
- maintain a realistic charge distribution between adjacent atomic centres in the molecule or ion.

**Problem set 5.9**

(a)
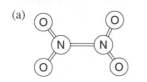

(b)
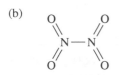

**5.9** (a) The planar structure of $N_2O_4$ — all nitrogen-oxygen bonds are equivalent; (b) a resonance structure to consider for question 6.

1. Draw resonance structures to describe the bonding in $NH_3$. Do all the structures contribute equally?
2. Draw resonance structures for $[NO_3]^-$. [*Hint*: Does N obey the octet rule?]
3. Provide a VB picture of the bonding in $[PO_4]^{3-}$ which is consistent with the fact that the P–O bonds are all the same length.
4. Draw a set of resonance structures for $PF_5$, indicating only those structures that contribute significantly to the bonding.
5. The azide ion, $[N_3]^-$, is linear. How is the bonding in this ion best described in terms of resonance structures? Is more than one structure needed to give a picture of the bonding that is consistent with the fact that the N–N bond distances are equal?
6. Figure 5.9a shows the structure of $N_2O_4$. Is Figure 5.9b a reasonable representation of the bonding in $N_2O_4$? Explain your answer.

## HYBRIDIZATION OF ATOMIC ORBITALS

The emphasis in this set of problems is for you to recognize appropriate hybridization schemes for atomic centres, and to be able to describe the bonding in a molecule or ion in terms of a hybridization scheme.

**Problem set 5.10**

**(5.5)**

1. What hybridization scheme would you assign to the carbon centre(s) in each of the following molecules (a) $CBr_4$; (b) $CO_2$; (c) $C_2H_4$; (d) $C_2H_2$; (e) $CH_2Cl_2$? On what have you based your decisions?
2. What atomic orbitals combine to give an $sp^2$ hybrid set of orbitals?
3. Why can double bond character be described in terms of an $sp^2$ hybridization scheme but is not compatible with an $sp^3$ model?
4. Describe the bonding in $[H_3O]^+$ in terms of a hybrid orbital model. What is the bond order of each O–H bond? In which orbital is the oxygen lone pair accommodated?
5. Describe the bonding in methanal, **5.5**, in terms of a hybrid orbital model.
6. Assign an appropriate hybridization scheme to the central atom in each of the following molecules or ions (a) $PF_5$; (b) $SF_6$; (c) $PF_3$; (d) $BF_3$; (e) $IF_5$; (f) $ClF_3$? [*Hint*: Use the VSEPR model first to assign a structure.]

# 6 Ions

---

**Topics**

- Ionization energies
- Electron affinities
- Ionic lattices
- Sizes of ions
- Determining the Avogadro constant from an ionic lattice
- Lattice energy

---

## IONIZATION ENERGIES

➤

**Values of ionization energies for use in this section may be found in Appendix 8 of H&C**

The first ionization energy ($IE_1$) of a gaseous atom is the internal energy change at 0 K associated with the removal of the first valence electron:

$$X(g) \rightarrow X^+(g) + e^-$$

The notation for this internal energy change is $\Delta U(0\ K)$, but for a thermochemical cycle we use a *change in enthalpy* and make the assumption that $\Delta H(298\ K) \approx \Delta U(0\ K)$. The validity of this assumption is discussed in the main text.

The second ionization energy ($IE_2$) of a gaseous atom refers to the process:

$$X^+(g) \rightarrow X^{2+}(g) + e^-$$

and the third ionization energy ($IE_3$) of a gaseous atom refers to the process:

$$X^{2+}(g) \rightarrow X^{3+}(g) + e^-$$

and so on. Note that each equation refers to an *ionization in the gas phase*.

**Problem set 6.1**

1. How are the values of $IE_2$ for Mg and $IE_1$ for Mg⁺ related?
2. Do you expect the value of $IE_1$ for Ne ($Z = 10$) to be greater than, less than, or about the same as the value of $IE_1$ for Na ($Z = 11$)? Give reasons for your answer. [Think about it before you look up the values!]
3. Does it require more, less or about the same amount of energy to remove an electron from an atom which possesses a $[He]2s^1$ configuration compared with that with an $[Ar]4s^1$ ground state configuration?
4. Using appropriate data from the appendices in H&C, determine the enthalpy change for the reaction:

$$Ca(s) \rightarrow Ca^{2+}(g) + 2e^-$$

5. The ionization of a metal atom to an ion $M^{n+}$ is always a highly endothermic process. Why, therefore, is the formation of many ionic salts of metal M thermodynamically favourable?

6. Figure 6.1 shows the trend in values of the first ionization energies ($IE_1$) from $Z = 1$ to 10. Give reasons for the following trends in the graph: (a) the increase from H to He; (b) the decrease from He to Li; (c) the decrease from N to O; (d) the increase from O to F to Ne; (e) the increase from Li to Be; (f) the decrease from Be to B; (g) the fact that $IE_1$(Li) is less than $IE_1$(H).

7. Figure 6.2 shows the trend in the values of the first five ionization energies of a metal X. Account for the trend.

8. From your answer to question 7, what do you expect will be the formula of (a) the chloride, (b) the oxide, and (c) the hydroxide formed by metal X?

**6.1** The trend in values of the first ionization energies ($IE_1$) for elements from $Z = 1$ to 10.

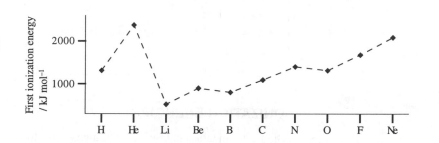

## ELECTRON AFFINITIES

➤
***Caution!***
**The sign convention for electron affinities is often a source of confusion and can lead to error**

The first electron affinity ($EA_1$) of an atom is *minus* the internal energy change at 0 K associated with the gain of one electron by a gaseous atom:

$$Y(g) + e^- \rightarrow Y^-(g)$$

For the most part, our use of electron affinities will be confined to that in thermochemical cycles and we define an *enthalpy* value $\Delta_{EA}H(298\ K)$ such that:

$$\Delta_{EA}H(298\ K) \approx \Delta_{EA}U(0\ K) = -EA$$

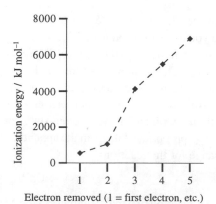

**6.2** For question 6, problem set 6.1.

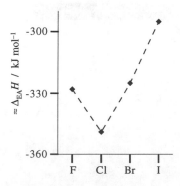

**6.3** Approximate (see text) values of $\Delta_{EA}H$ for the group 17 elements.

**Problem set 6.2**

1. Explain why the attachment of an electron to a fluorine atom is an exothermic process.
2. Why is the attachment of an electron to the $O^-$ ion an endothermic process?
3. Since it appears to be unfavourable to form $O^{2-}$ from $O^-$, why are many metal oxides thermodynamically stable?
4. Figure 6.3 shows the trend in the values of $\Delta_{EA}H$ for the first four elements in group 17. Discuss the factors that contribute to this trend.
5. Using appropriate data from the appendices in H&C, determine the enthalpy change associated with the reaction:

$$O_2(g) + 4e^- \rightarrow 2O^{2-}(g)$$

## IONIC LATTICES

In H&C, we described several lattice types and in this section we combine questions regarding these structures with some which introduce new lattice types. One point that we emphasize is the relationship between structure and compound stoichiometry. You will find full descriptions of the NaCl (rock salt), CsCl, $CaF_2$ (fluorite), $TiO_2$ (rutile) and ZnS (zinc blende and wurtzite) lattices in H&C, Sections 6.8 to 6.13.

**Problem set 6.3**

1. What is meant by the term *unit cell*.
2. The right-hand diagram in Figure 6.4 corresponds to the unit cell of CsCl that we described in the main text. Why is the left-hand diagram in Figure 6.4 an equally valid representation of the unit cell?
3. Why is it not possible to represent a unit cell of the NaCl lattice using the diagram in Figure 6.5? Why must the unit cell contain eight of these cubic-motifs (see Figure 6.10, p. 55).
4. A bromide of indium, $InBr_x$, crystallizes in the NaCl lattice structure but only one third of the metal ion sites are occupied. What is the oxidation state of indium in this compound?
5. Cerium(IV) oxide is a catalyst in self-cleaning ovens. In its crystalline state, the cerium(IV) oxide lattice contains four-coordinate oxygen centres. (a) What is the coordination number of each cerium(IV) centre? (b) Cerium(IV) oxide adopts one of the lattice types described in H&C; which one is it?

 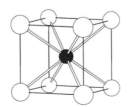                      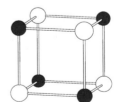

**6.4** The two possible unit cells of the CsCl lattice.

$\bigcirc$ = Cl    ● = Cs

**6.5** For question 3, problem set 6.3.

$\bigcirc$ = Cl    ● = Na

6.    Gallium phosphide is used in solar energy conversion. It crystallizes with the zinc blende lattice. (a) What is the formula of gallium phosphide? (b) Is this formula consistent with the positions of gallium and phosphorus in the periodic table?

7.    The unit cell of iridium dioxide is shown in Figure 6.6. (a) Confirm that the ratio of Ir : O atoms in the compound is 1 : 2. (b) Which lattice-type is this?

In this next set of problems, we apply the knowledge learnt about representative lattice-types to some new situations. Firstly, remind yourself about the way in which atoms are shared between unit cells. Fill in the gaps in the following statements:

In a cubic or cuboid unit cell, an ion in a corner site is shared between ___ unit cells, an ion on a face is shared between ___ unit cells, and an ion on an edge is shared between ___ unit cells.

In a unit cell consisting of a hexagonal prism (structure **6.1**), an ion in a corner site is shared between ___ unit cells, an ion on a face is shared between ___ unit cells, and an ion on a vertical edge is shared between ___ unit cells.

**(6.1)**

**Problem set 6.4**

1.    Figure 6.7 shows the arrangement of ions in a unit cell of niobium oxide. (a) What is the coordination number of each oxygen and niobium centre? (b) In what structural environment is each oxygen and niobium centre? (c) Determine the stoichiometry of niobium oxide. (d) Often ionic lattices are described in terms of prototype lattices which include those described in H&C. To which of these lattices is that of niobium oxide related? *Write* a clear description to illustrate how the two lattices are related that would allow someone who knew the relevant prototype lattice to draw the unit cell of niobium oxide. [*Hint*: Draw in the edges that define the cubic unit cell in Figure 6.7.]

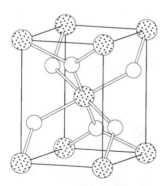

**6.6**  A unit cell of iridium dioxide.

= Ir    ◯ = O

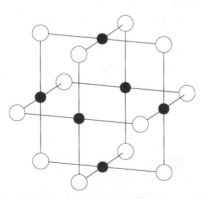

**6.7**  The ions in a unit cell of niobium oxide.

● = Nb  ◯ = O

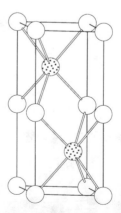

**6.8**  A unit cell of nickel arsenide.

◯ = Ni    = As

2.  A unit cell of the nickel arsenide lattice is shown in Figure 6.8. This lattice type is adopted by a range of compounds including an iron tin alloy that is used for tinplate (e.g. in tin cans). (a) Determine the formula of nickel arsenide from the unit cell, and hence determine the formula of the iron tin alloy. (b) What is coordination number of each nickel centre in nickel arsenide? (c) What is the coordination environment around each arsenic centre? [*Hint*: Use Appendix 15 in H&C to help you.]

3.  Crystalline $K_2O$ and $K_2S$ each adopt an *antifluorite* lattice. Use your knowledge of the fluorite lattice to deduce what is meant by an antifluorite lattice-type.

4.  The oxide $FeSb_2O_6$ adopts a *trirutile* lattice, a unit cell of which is shown in Figure 6.9. (a) By drawing in appropriate cell edges, illustrate how this unit cell is related to three unit cells of the rutile lattice-type. (b) Why can the solid state structure of $FeSb_2O_6$ not be described in terms of a single unit cell of the rutile lattice? (c) What is the coordination environment of each atom type? (d) By using the information in the unit cell, confirm the stoichiometry of this iron antimony oxide. (e) What oxidation states would you assign to each element in $FeSb_2O_6$?

**6.9** The trirutile lattice adopted by $FeSb_2O_6$.

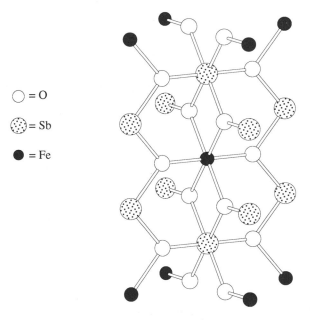

$\bigcirc$ = O

$\circ\!\!\cdot\!\!\circ$ = Sb

● = Fe

**See also Box 7.1 in H&C**

➤ Ionic lattices may be viewed in terms of the close packing of spherical anions with cations occupying the spaces (interstitial holes) between the anions. We look at this approach in Chapter 7 of the workbook.

## SIZES OF IONS

The radii of anions and cations can be estimated from the internuclear distances in an ionic lattice (equation 6.1).

$$\text{Internuclear distance between a cation and the closest anion in a lattice} = r_{cation} + r_{anion} \qquad (6.1)$$

**Problem set 6.5**

1.  How are ionic radii estimated experimentally? Point out the assumptions that are made during the estimations and comment on their validity.
2.  (a) Using the data in Table 6.1, can you obtain a value for the ionic radius of a $K^+$ ion? State clearly any approximations that you make. (b) Given a value of $r_{Na^+}$ of 102 pm, give the best estimate of $r_{K^+}$.
3.  Do the ionic radii of the halide ions (a) increase, (b) decrease, or (c) stay about the same on descending group 17?
4.  How does the radius of an oxygen atom compare with that of an $O^{2-}$ ion? Give reasons for your answer.
5.  How does the radius of a calcium atom compare with that of a $Ca^{2+}$ ion? Give reasons for your answer.
6.  The ionic radii of six-coordinate $Mn^{2+}$, $Mn^{3+}$ and $Mn^{4+}$ are 67, 58 and 39 pm respectively. Account for this trend.

**Table 6.1**   Data for question 2, problem set 6.5.

| Compound | Lattice type | Internuclear distance between cation and closest anion / pm |
|----------|--------------|-----------------------------------------------------------|
| NaF      | NaCl         | 231                                                       |
| NaCl     | NaCl         | 281                                                       |
| KF       | NaCl         | 266                                                       |
| KCl      | NaCl         | 314                                                       |
| RbF      | NaCl         | 282                                                       |
| RbCl     | NaCl         | 329                                                       |

## DETERMINING THE AVOGADRO CONSTANT FROM AN IONIC LATTICE

The Avogadro number can be estimated using X-ray diffraction data for a crystal lattice. The example chosen is the NaCl lattice, a unit cell of which is shown in Figure 6.10. The calculation uses the following data:

➤
**X-ray diffraction: see Section 3.2 in H&C**

*   X-ray diffraction results give the internuclear Na–Cl separation as 281 pm;
*   the density of sodium chloride is 2.165 g cm$^{-3}$;
*   values of $A_r$ for Na and Cl are 23 and 35.5 respectively.

**6.10**  A unit cell of the NaCl lattice. *Question*: This unit cell has chloride ions occupying the corner sites. Compare this with Figure 6.11 in H&C where the unit cell showed sodium ions in the corner sites. The two unit cells are equally valid. Why is this?

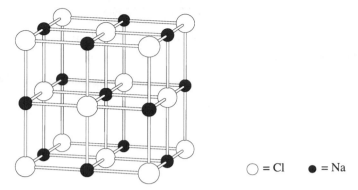

$\bigcirc$ = Cl    $\bullet$ = Na

The unit cell contains four NaCl ion-pairs; you should confirm this by considering the sharing of ions between adjacent unit cells.

The length of each side of the unit cell is *twice* the internuclear Na—Cl separation (equation 6.2) and the volume of the unit cell is given by equation 6.3.

$$\text{Length of one side of the unit cell} = 2 \times 281 = 562 \text{ pm} \tag{6.2}$$

$$\text{Volume of the unit cell} = \left(\text{Length of side of unit cell}\right)^3 \tag{6.3}$$

$$= 1.775 \times 10^8 \text{ pm}^3$$

The volume of one ion-pair is one quarter of the volume of the unit cell, and since one mole of sodium chloride contains the Avogadro number, $L$, of ion-pairs, the volume occupied by a mole of sodium chloride is given by equation 6.4.

$$\text{Volume of one mole of NaCl} = \frac{1.775 \times 10^8 \times L}{4} \text{ pm}^3 \tag{6.4}$$

We can also express the volume of one mole of NaCl in terms of the relative molecular mass and the density (equation 6.5).

$$\text{Volume of one mole of NaCl} = \frac{\text{Relative molecular mass}}{\text{Density}} \tag{6.5}$$

$$= \frac{58.5}{2.165} = 27.0 \text{ cm}^3$$

Hence, equations 6.6 equate the two expressions for the volume of one mole of NaCl and hence we can find the Avogadro constant. Notice that before we can write equations 6.6, we must ensure that the units of volume from equations 6.4 and 6.5 have been made consistent. (What conversion has been applied?)

$$\left.\begin{array}{l} \dfrac{1.775 \times 10^{-22} \times L}{4} = 27.0 \\[2em] L = \dfrac{27.0 \times 4}{1.775 \times 10^{-22}} = 6.09 \times 10^{23} \text{ mol}^{-1} \end{array}\right\} \tag{6.6}$$

---

## LATTICE ENERGY

### The Born-Landé equation

**Problem set 6.6**

1.    Write down the Born-Landé equation. Which part of the expression deals with (a) attractive interactions and (b) repulsive interactions? What do you understand by the Madelung constant, $A$, and the Born exponent, $n$?
2.    Is the value of the lattice energy derived from the Born-Landé equation an internal energy or an enthalpy value?
3.    Use Table 6.8 (p. 286) in H&C to determine an appropriate value of the Born exponent for (a) RbF; (b) LiCl; (c) MgO; (d) ZnS; (e) CaO.
4.    Using the Born-Landé equation, determine the lattice energy of RbF; it has a sodium chloride lattice-type. The Born exponent was found in question 3. Other data required: $A = 1.7476$; $L = 6.022 \times 10^{23}$ mol$^{-1}$; internuclear Rb–F separation = 282 pm; $\varepsilon_0 = 8.854 \times 10^{-12}$ F m$^{-1}$; $e = 1.602 \times 10^{-19}$ C.

### The Born-Haber cycle

**Problem set 6.7**

1.    For what chemical reaction is the lattice energy of magnesium chloride defined?
2.    Construct a Born-Haber cycle that may be used to estimate the lattice energy of magnesium chloride.
3.    Is the value of the lattice energy derived from a Born-Haber cycle an internal energy or an enthalpy value? Are the two quantities similar in magnitude?
4.    Estimate the lattice energies of (a) calcium fluoride and (b) potassium oxide using Born-Haber cycles and appropriate data from the appendices in H&C.

**7** | # Elements

---

**Topics**
- Packing of spheres
- Packing of spherical ions: ionic lattices
- The dependence of physical properties of elements on structure
- Resistance and conductivity: Ohm's Law
- Metallic bonding and semiconductors

---

## PACKING OF SPHERES

The ideas of 'packing of spheres' are used when we describe the solid state structures of, for example, the group 18 elements, dihydrogen, difluorine, metals and some alloys, and can also be used to describe some ionic lattices (see p. 59 of the workbook). You should ensure that you are familiar with the following types of packing of spheres:

**ccp = cubic close-packed**
**hcp = hexagonal close-packed**
**bcc = body-centred cubic**
**fcc = face-centred cubic**

- cubic close-packing (face-centred cubic lattice)
- hexagonal close-packing
- simple cubic lattice
- body-centred cubic lattice.

**Problem set 7.1**

1. Two unit cells are shown in Figure 7.1. What lattice types do they describe? How many nearest neighbours does a sphere in each unit cell possess?
2. If metal atoms were to adopt a lattice with the unit cell shown in Figure 7.1a, would the packing of the atoms be more or less efficient than that in a lattice with the unit cell shown in Figure 7.1b?
3. What is the short-hand notation used to denote the sequence of layers in (a) a hexagonal close-packed assembly and (b) a cubic close-packed assembly?
4. How many nearest neighbours does a sphere in (a) a hexagonal and (b) a cubic close-packed assembly possess?

**7.1** Unit cells for questions 1 and 2, problem set 7.1.

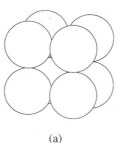

(a)

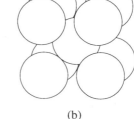

(b)

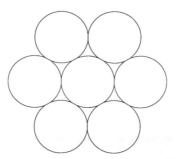 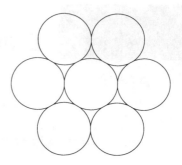

**7.2** Part of a layer of spheres in a close-packed assembly for question 5, problem set 7.1.

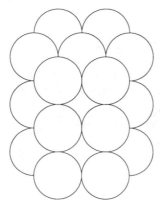

**7.3** An assembly of spheres for question 6 in problem set 7.1.

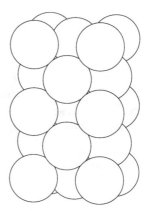

**7.4** Close-packed assembly for question 7 in problem set 7.1.

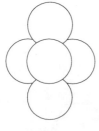

**7.5** Arrangment of five spheres for question 8, problem set 7.1.

5. Figure 7.2 shows two identical representations of part of a layer of a close-packed assembly of spheres. (a) On the left-hand diagram, mark the positions occupied by the second layer of atoms in a hexagonal close-packed assembly. (b) Do these positions differ from those occupied by the second layer of atoms in a cubic close-packed assembly? (c) Assuming that the sites you have chosen in the left-hand diagram are *now occupied*, mark on the right-hand diagram the positions occupied by the *third* layer of atoms in a hexagonal close-packed assembly.

6. Figure 7.3 shows spheres in one of the assemblies listed at the beginning of the section; the figure does *not* show a unit cell. This view of the assembly shows *square-faces* which are a particular feature of this mode of packing. (a) Mark in the square-faces in the figure. (b) Which type of packing of spheres is illustrated in Figure 7.3? (c) What type of interstitial holes are present in this assembly? (d) Indicate in the Figure 7.3 where the interstitial holes are found. [*Hint:* consider how the assembly is further extended.]

7.    Figure 7.4 shows part of a close-packed assembly of spheres. (a) Which type of assembly is shown? (b) How many unit cells are shown in the figure? (c) Mark on the diagram a set of spheres that define *one* unit cell. (d) What is the alternative name for this type of packing?

8.    Would the unit of five spheres shown in Figure 7.5 be present in (a) a ccp, (b) an hcp, (c) a simple cubic, or (d) a bcc assembly?

## PACKING OF SPHERICAL IONS: IONIC LATTICES

In Box 7.1 of H&C, we introduced the idea that in an ionic compound it is often the case that the anions are significantly larger than the cations and that the lattice structure may be considered in terms of 'small' cations occupying the interstitial holes between 'large' anions. In the case of zinc blende, we saw that the $S^{2-}$ ions ($r_{ion}$ = 184 pm) adopt an fcc arrangement and the $Zn^{2+}$ ions ($r_{ion}$ = 60 pm) occupy the tetrahedral holes in the close-packed array. We now consider the sodium chloride lattice in the same way.

### The NaCl-type lattice

Figure 7.6a shows a *space-filling* diagram of a unit cell of an alkali metal halide MX which adopts the NaCl-lattice. The $X^-$ anions adopt a fcc arrangement. In the close-packed sphere model, we assume that the larger anions touch each other and that the smaller cations fit neatly into the interstitial holes. The problem set below guides you through a critical analysis of this model.

**Problem set 7.2**    1.    Figure 7.6b shows one face of the unit cell of the alkali metal halide MX drawn in Figure 7.6a; assume that the $M^+$ and $X^-$ ions are spherical and that they touch one another. Derive a relationship between the radius of the cation ($r_{M^+}$) and the radius of the anion ($r_{X^-}$). This relationship applies to an ideal situation in which cations occupy the octahedral holes in the fcc arrangement of $X^-$ ions.

**7.6** (a) A space-filling diagram of a unit cell of a group 1 metal halide MX which adopts the NaCl-lattice; (b) one face of the same unit cell.

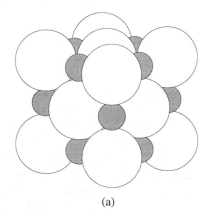

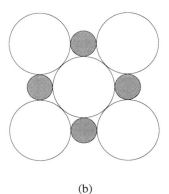

(a)                                            (b)

**Table 7.1**  Some compounds which adopt the NaCl lattice.

| Compound | Radius of cation / pm | Radius of anion / pm |
|----------|-----------------------|----------------------|
| LiF | 76 | 133 |
| NaF | 102 | 133 |
| KF | 138 | 133 |
| CsF | 170 | 133 |
| NaCl | 102 | 181 |
| KCl | 138 | 181 |
| LiBr | 76 | 196 |
| LiI | 76 | 220 |

2.   All the alkali metal halides listed in Table 7.1 adopt the NaCl-type lattice. Use the data in the table and your answer to question 1 to critically assess how realistic the close-packed sphere model is for this set of compounds.

3.   Which compounds in Table 7.1 appear to be the 'odd ones out' and why? Why is it possible to still refer to these compounds as having NaCl-type structures?

## THE DEPENDENCE OF PHYSICAL PROPERTIES OF ELEMENTS ON STRUCTURE

**Problem set 7.3**

**For this set of questions, you will need to refer to appropriate data in Appendices 10 and 11 in H&C**

1.   Describe what happens when each of the following elements melts (a) copper; (b) hydrogen; (c) bromine; (d) orthorhombic sulfur; (e) argon; (f) sodium. How do the *relative values* of the melting points of these elements relate to the processes you have described?

2.   Write down equations to define the processes for which the standard enthalpies of atomization of (a) chlorine, (b) zinc, (c) white phosphorus, and (d) potassium refer?

3.   What is the relationship between the values of $D(F–F)$ and $\Delta_a H°$ of fluorine?

4.   (a) Why do the alkali metals tend to exhibit lower values of $\Delta_a H°(298 \text{ K})$ than metals from the $d$-block? (b) Give one $d$-block metal which is an exception and suggest a reason for this observation.

5.   What is the periodic relationship between boron and aluminium? The melting points of β-rhombohedral boron and aluminium are 2453 K and 933K. Suggest reasons for this dramatic difference.

6.   To what processes do the values of (a) $\Delta_{fus} H = 0.7$ kJ mol$^{-1}$, (b) $\Delta_{vap} H = 5.6$ kJ mol$^{-1}$ and (c) $\Delta_a H° = 473$ kJ mol$^{-1}$ for nitrogen refer? Are these data tabulated per mole of $N_2$ molecules or per mole of N atoms?

7.   Figure 7.7 shows the trends in values of $\Delta_a H°(298 \text{ K})$ and melting points for

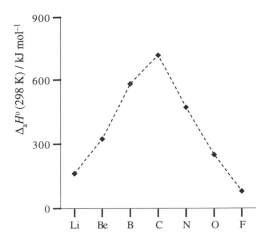

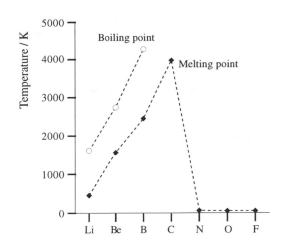

**7.7** Trends in (a) $\Delta_a H^o$(298 K) and (b) melting and boiling points for elements with atomic numbers between 3 and 9.

Chromatography is a *separation* technique, one type of which is *column chromatography*. The components to be separated are *adsorbed* on to a *stationary phase* such as alumina or silica, and then *eluted* with a suitable solvent or solvent-mixture. Some components are removed from the stationary phase more readily than others and separation of the mixture is thereby achieved.

Gas chromatography: see Box 8.3 in H&C

the elements from $Z = 3$ to 9. Figure 7.7b also shows the values of the boiling points for lithium, beryllium and boron; carbon sublimes and were the boiling points for the last three elements to be plotted, the values would almost superimpose those of the melting points. (a) Why do nitrogen, oxygen and fluorine have short liquid ranges? (b) Account for the trend in the values of $\Delta_a H^o$(298 K). (c) Why is there a dramatic decrease in the melting point in going from carbon to nitrogen? (d) What factors contribute towards the increase in melting points from Li to C?

8. The densities of the allotropes $\alpha$-graphite and diamond are 2.3 and 3.5 g cm$^{-3}$. Account for this difference.

9. The allotropes $C_{60}$ and $C_{70}$ can be separated from each other by using *column chromatography*, but graphite cannot be separated from diamond by this technique. Explain what physical properties of the fullerenes make them so different from the other allotropes of carbon. What property in particular is needed for column chromatography to be a successful technique?

## RESISTANCE AND CONDUCTIVITY: OHM'S LAW

In this section, we illustrate applications of Ohm's Law which states that the 'current passing through a wire at constant temperature is proportional to the potential difference between the ends of the wire'. Ohm's Law may be expressed in the form of equation 7.1 where $V$ is the potential difference in volts (V), $I$ is the current in amperes (A), and $R$ is the resistance in ohms ($\Omega$).

$$V = I \times R \qquad \qquad Ohm's \ Law \qquad \qquad (7.1)$$

Ohm's law can be applied to the electrical conductivity of metal and non-metal wires. *Resistance is dependent upon temperature.* In H&C we discussed the resistivity of some elements: the *electrical resistivity* ($\rho$) gives the resistance of a wire as a function of its length and cross section (equation 7.2). The units of resistivity are $\Omega$ m (*not* $\Omega$ m$^{-1}$ or $\Omega$ m$^{-3}$).

$$\text{Resistance (in } \Omega) = \frac{\text{Resistivity (in } \Omega \text{ m}) \times \text{Length of wire (in m)}}{\text{Cross section (in m}^2)}$$

$$R = \frac{\rho \times l}{a} \tag{7.2}$$

**Problem set 7.4**

1. A potential difference of 12 V is applied across a 4.6 $\Omega$ resistor. Determine the current flowing.
2. The electrical resistivity of zinc in 6 $\mu\Omega$ cm. What is this in $\Omega$ m?
3. The electrical resistivity of titanium is $4.3 \times 10^{-7}$ $\Omega$ m. What will be the resistance of a 0.5 m piece of titanium wire with cross section $8 \times 10^{-7}$ m$^2$?
4. If a potential difference of 6 V is applied across the ends of the wire in question 3, what current will flow along the wire? Assuming that the wire has a circular cross section, would an increase in its diameter cause an increase or decrease to the amount of current flowing?
5. Account for the fact that the electrical resistivity of a rod of $\alpha$-graphite is direction dependent.
6. Using the data in Figure 7.8, estimate the resistance that a 0.2 m copper rod of cross section $1 \times 10^{-4}$ m$^2$ would exhibit at (a) 298 K and (b) 500 K.

**7.8** Variation in electrical resistivity of a copper wire with temperature. Notice the way in which the resistivity scale is denoted: it runs from 0 to $3.5 \times 10^{-8}$ $\Omega$ m.

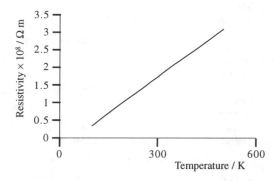

## METALLIC BONDING AND SEMICONDUCTORS

**Problem set 7.5**

1. Give a brief explanation of what is meant by *band theory* with reference to the bonding in lithium.
2. What is a *band gap*?
3. Which of the following would you associate with (a) a metal, (b) a semiconductor, and (c) an insulator? (i) A large band gap between filled and vacant bands; (ii) a very small band gap between filled and vacant bands; (iii) a partially occupied band.

➤
**In H&C, Section 7.13, we discussed semiconductors only in general terms**

4. Two types of semiconductors, *n*- and *p*-types, are made by *doping* a host with atoms containing more or less valence electrons, respectively, than atoms of the host. How do you think doping silicon with arsenic atoms (an atom substitution) to give an *n*-type semicondutor would change the electrical conductivity of the material?

# 8 Alkanes, alkenes and alkynes

**Topics**
- Nomenclature and isomerism
- Conformation
- Chiral compounds
- Reactions of alkanes
- Reactions of alkenes
- A study of *syn-* and *anti-*addition
- Addition polymers
- Reactions of alkynes
- Optical activity and chirality

## NOMENCLATURE AND ISOMERISM

### Fundamental points

This next problem set tests your knowledge of some basic facts about alkanes, alkenes and alkynes.

**Problem set 8.1**

1. What do the terms *hydrocarbon*, *aliphatic*, *acyclic* and *cyclic* mean?
2. An alkane does *not* possess a functional group. What is a *functional group*?
3. What is the functional group characterising (a) an alkene, and (b) an alkyne? Are these types of compound saturated or unsaturated?
4. In the following hydrocarbons, state the type of hybridization that is appropriate for each carbon centre:

   $CH_3CH_2CH_2CH_3$     $CH_3CH=C=C=CH_2$     $CH_2=CHCH_3$     $CH_3C\equiv CCH_3$

### Chemical formulae

The chemical formulae of compounds (not only organic compounds) can be written in different ways. In H&C we introduced *empirical* and *molecular formulae* in Section 1.23, and also said that *structural formulae* provided information about the way in which atoms were connected. Throughout *Chemistry: An Integrated Approach* we use various types of formulae to convey structural details. The names given to these formulae are listed in Table 8.1 using propene as the example.

### Simplifying structural diagrams

**(8.1)**

It is often useful to simplify displayed and stereochemical formulae by omitting hydrogen atoms. If we draw the structure of butane as shown in **8.1**, it is assumed

**Table 8.1**  Types of chemical formulae, using propene as the example

| Formula name | Information provided | Example: Propene |
|---|---|---|
| Empirical | Stoichiometric ratio | $CH_2$ |
| Molecular | Formula consistent with $M_r$ | $C_3H_6$ |
| Structural | Connectivity of atoms | $CH_3CH=CH_2$ |
| Displayed | Projection of atoms and bonds | 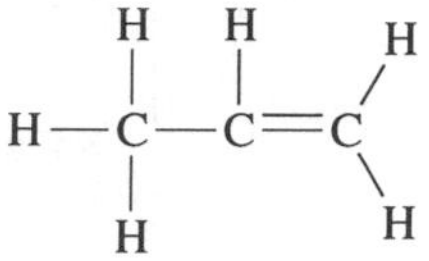 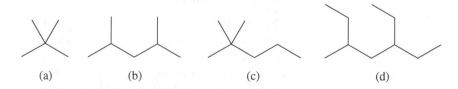 |
| Stereochemical | Stereochemistry | 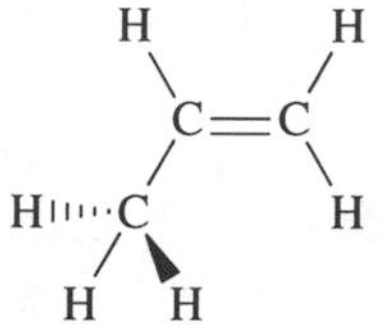 |

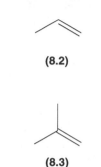

**(8.2)**

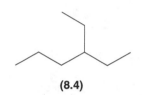

**(8.3)**

**(8.4)**

that the connectivity of each carbon atom is made up to four by the presence of an appropriate number of H atoms. Structure **8.2** shows a shorthand way of drawing the structure of propene. Lines drawn as side-chains, for example in structure **8.3** signify the presence of methyl groups; in structure **8.4**, the side-chain is an ethyl group. Use this method of drawing displayed formulae wherever possible.

## Alkanes

**Problem set 8.2**

1. What is the general formula of a straight chain alkane?
2. Write down the formulae of (a) butane, (b) octane, (c) decane and (d) hexane.
3. Name the following alkanes.

    (a)        (b)        (c)        (d)

4. Draw structures for the alkanes: (a) 2,5-dimethylheptane; (b) 2,3,4-trimethyloctane; (c) 3,3,4,4-tetraethyldecane.
5. How many acyclic isomers does $C_5H_{12}$ possess? What are their structures?
6. Draw structural formulae for (a) *tertiary*-butyl and (b) isopropyl groups.

## Alkenes

1. What is the general formula of a straight chain alkene (with one double bond)?
2. Write down the structural formulae of (a) but-1-ene, (b) pent-2-ene, (c) buta-1,3-diene and (d) hexa-2,4-diene.
3. Name the following alkenes: (a) $CH_3CH_2CH_2CH_2CH=CH_2$; (b) $CH_3CH=C=CH_2$; (c) $CH_3CH_2CH=CHCH_2CH_3$.
4. What do the prefixes (E)- and (Z)- denote? Use but-2-ene to exemplify your answer.
5. How many acyclic isomers (including structural and geometrical) does $C_5H_{10}$ possess? Draw the structure of each compound.

## Alkynes

1. What is the geometry about a carbon atom involved in a C≡C triple bond?
2. What are the structures of (a) but-2-yne, (b) hex-1-yne and (c) penta-1,4-diyne.
3. Does butyne possess structural isomers?
4. Does but-2-yne possess geometrical isomers?

## CONFORMATION

Although we refer to 'straight chain' alkanes, this does not mean that the carbon backbone is actually 'straight', merely 'unbranched'. Rotation occurs about single bonds. Since C–H bonds are *terminal*, rotation about them does not alter the *conformation* of the carbon chain in a hydrocarbon. Rotation about a C–C bond may alter the conformation. *Conformers* possess different *steric energies* arising from van der Waals interactions between the hydrogen atoms. A conformer may be energetically unfavourable if the hydrogen atoms approach too closely. In ethane, the *staggered conformer* is at a lower energy (more favoured) than the *eclipsed conformer*. *Skew conformers* lie in between the eclipsed and staggered conformations.

➤ Staggered and eclipsed conformers of ethane: see H&C, Figure 8.9

**8.1**   Conformers of octane.

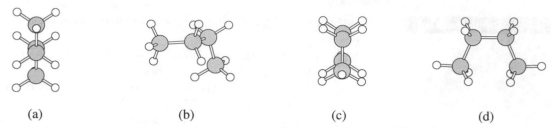

(a)          (b)          (c)          (d)

**8.2**   Conformers of butane related by rotation about the C(2)–C(3) bond.

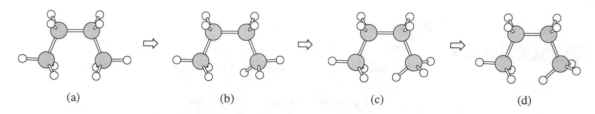

(a)          (b)          (c)          (d)

**8.3**   Conformers of butane related by rotation about the C(1)–C(2) bond.

**Problem set 8.5**

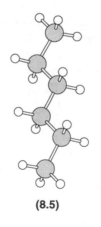

**(8.5)**

1.   Figure 8.1 shows three conformers of octane. The second conformer is generated from the first by rotation about *one* C–C bond, and the third conformer is generated from the second by rotation about a different C–C bond. About which bonds does rotation occur?

2.   Figure 8.2 shows ball-and-stick diagrams of butane in various conformations. Label each appropriately as being a staggered, eclipsed or skew conformer.

3.   Refer to Figure 8.3. (a) As you move from structure (a) to (d), about which bond is rotation occurring, and in which direction? (b) Why is the conformation of the CCCC-backbone unchanged by this bond rotation? (c) How do the inter-hydrogen interactions alter in going from (a) to (d)?

4.   Which of the following compounds possess conformers? (a) ethene; (b) hex-1-ene; (c) hexa-1,5-diene; (d) propyne. Rationalize your answer.

5.   How would you describe the conformation of the alkane shown in structure **8.5**?

## CHIRAL COMPOUNDS

The aims of the problems in this section are to:
  • be able to recognize whether a compound is chiral or achiral,
  • be able to find asymmetric carbon centres in an organic molecule, and
  • realize that the presence of asymmetric carbon atoms does not *necessarily* mean that a compound is chiral.

**1.** Point out the asymmetric carbon centres in the following compounds; not all the molecules contain chiral centres.

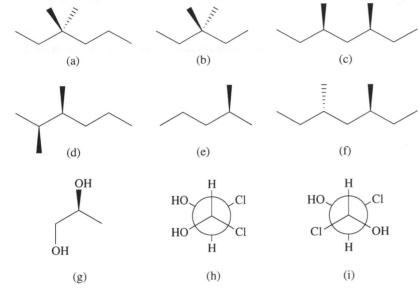

(a)        (b)        (c)

(d)        (e)        (f)

(g)        (h)        (i)

> **Enantiomers are also known as optical isomers because they interact with polarized light in opposite ways.**

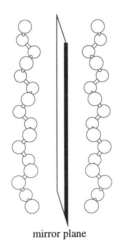

mirror plane

**8.4** Two chains of *catena*-sulfur.

**2.** Which of the compounds in question 1 possess planes of symmetry? Is it necessary to provide stereochemical formulae in question 1, or would formulae of the type shown in structure **8.4** have been sufficient?

**3.** Which of the compounds in question 1 are chiral? Does the presence of an asymmetric carbon atom necessarily mean that a compound is chiral?

**4.** Figure 8.4 shows two chains of *catena*-sulfur, $S_\infty$. Is *catena*-sulfur chiral.

## REACTIONS OF ALKANES

### Combustion of alkanes

**1.** Determine the standard enthalpy of combustion of octane if $\Delta_f H^\circ(H_2O, l) = -286$; $\Delta_f H^\circ(CO_2, g) = -393.5$; $\Delta_f H^\circ(C_8H_{18}, g) = -250$ kJ mol$^{-1}$.

**2.** What volume of $CO_2$ is produced (at 300 K and 1 bar pressure) from the complete combustion of 3 moles of octane? [Volume of 1 mole of gas at 273 K and 1 bar = 22.7 dm$^3$.]

**3.** After complete combustion, 0.25 moles of an acyclic alkane **X** produce 66 g $CO_2$ and 31.5 g $H_2O$. Suggest a possible identity for **X**, and draw displayed formulae for all possible isomers of **X**.

**4.** The standard enthalpies of combustion of fuels may be measured using bomb calorimetry, and these data may be used to determine standard enthalpies of formation. Calculate $\Delta_f H^\circ$ for butane if $\Delta_c H^\circ(C_4H_{10}, g) = -2878$ kJ mol$^{-1}$. [$\Delta_f H^\circ(H_2O, l) = -286$; $\Delta_f H^\circ(CO_2, g) = -393.5$ kJ mol$^{-1}$.]

## Halogenation of alkanes

**Problem set 8.8**

1.  (a) Suggest possible products when methane reacts with $Br_2$. (b) Why must the reactants be irridiated?
2.  (a) What *types* of carbon centres are present in butane? (b) What radicals could be formed by hydrogen abstraction? (c) Under appropriate conditions, would the formation of these radicals be equally favoured?
3.  Consider the reaction between 2-methylpropane and $Br_2$ under photolytic conditions. (a) Write an equation to show an initiation step. (b) Write equations to show possible propagation steps. (c) Of the hydrocarbon radicals that you show in the propagation steps, which is likely to be favoured and why? (d) Write equations to show possible termination steps.
4.  The bromination of methane under photolytic conditions is *inhibited* by the addition of HBr. Suggest a reason for this observation.
5.  During the chlorination of ethane, butane is a product. Account for this observation.

## REACTIONS OF ALKENES

### Addition reactions: observations

**Problem set 8.9**

1.  Give the reagents needed to achieve the following transformations.
    (a) $CH_2=CH_2 \rightarrow CH_2BrCH_2Br$
    (b) $CH_2=CH_2 \rightarrow CH_3CH_2Br$
    (c) $CH_3CH_2CH=CH_2 \rightarrow CH_3CH_2CH_2CH_3$
    (d) $CH_3CH=CH_2 \rightarrow CH_3CH(OH)CH_2(OH)$
    (e) $CH_3CH=CH_2 \rightarrow CH_3CHClCH_3$
    (f) $CH_3CH=CH_2 \rightarrow CH_3CH(OH)CH_3$
    (g) $CH_3CH=CH_2 \rightarrow CH_3CH_2CH_2OH$
2.  When ethene is bubbled through aqueous bromine water, what do you *observe*? Account for the observation.
3.  What products would be obtained from the reaction of hex-2-ene with ozone followed by reductive hydrolysis?
4.  What would be product of the reaction of 2-methylpropene with HCl?
5.  The addition of $Cl_2$ to an alkene gives a *vicinal dichloride*. What characterizes this type of compound?
6.  How might but-1-ene be converted to but-2-ene?

### Addition reactions: mechanisms

**Problem set 8.10**

1.  What feature of an alkene makes the compound susceptible to attack by electrophiles?
2.  Dichlorine is non-polar and yet it is able to provide a source of $Cl^+$ when it reacts with an alkene. Explain how this is possible.

3.  (a) Give a mechanism for the reaction of HBr with ethene. (b) Draw a reaction profile for this reaction and show the nature of the intermediate. (c) Which step is rate determining?

4.  What do you understand by the term *Markovnikov addition* to an alkene? Use the acid-catalysed addition of water to 2-methylpropene to illustrate Markovnikov addition, outlining the mechanism of the reaction.

5.  Classify the following carbenium ions as being primary, secondary or tertiary.

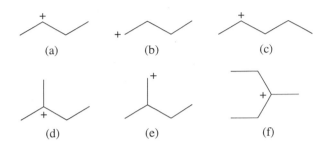

6.  The rate of addition of $H^+$ to a C=C bond depends on the stability of the carbenium ion formed in the rate determining step of the reaction. It follows that the observed *relative reactivities* of alkenes towards attack by $H^+$ depends upon the carbenium ion that is formed. Place the following alkenes in the expected order of reactivity towards the addition of $H^+$.

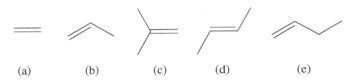

7.  Why is the addition of HCl to propene said to be *regioselective*?

8.  In what ways does the mechanism of the electrophilic addition of $Cl_2$ to an alkene differ from that of $Br_2$? What experimental evidence is there to support this proposal?

## A STUDY OF *SYN*- AND *ANTI*-ADDITION

➤ **See pages 374-375 of H&C**

➤ *Stereoisomers* differ only in the spatial orientations of atoms or groups in a molecule; optical isomers are a class of stereoisomers.

The *syn*- and *anti*-addition of HX to the general alkene shown in structure **8.6** gives rise to different *stereoisomers* of the product (Figure 8.5). Now we look more closely at these stereoisomers.

### The consequences of the addition product containing two asymmetric carbon atoms

You should notice in Figure 8.5 that the products of the *syn*- and *anti*-addition of HX to alkene **8.6** possess two asymmetric carbon atoms. Each compound is chiral

**8.5** *Syn-* and *anti*-addition of HX to an alkene ABC=CAB.

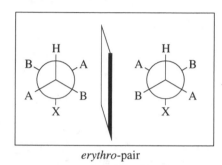

(8.6)

**8.6** The two pairs of enantiomers (drawn as Newman projections) formed from the *syn-* and *anti*-addition of HX to an alkene ABC=CAB.

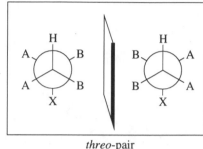

*erythro*-pair                    *threo*-pair

➤
**Newman projections: see Section 8.5 in H&C**

and consists of a pair of enantiomers. These are shown in Figure 8.6 along with the names that are used to distinguish them. These names may be assigned as follows:

- Draw Newman projections of the enantiomers of the addition product in *staggered* conformations; view the molecule along the C–C bond that was originally the C=C bond of the alkene.
- Ensure that the H and X groups from the reagent HX added to the alkene point up and down in the Newman projection.
- Two of the enantiomers will possess *identical* groups on the *same* side of the Newman projection and these are called the *threo*-pair of enantiomers.
- The other two enantiomers have *differerent* groups on the *same* side of the Newman projection and these are called the *erythro*-pair.

**Problem set 8.11**

1. Represent the four enantiomers in Figure 8.6 as sawhorse drawings.
2. Consider the addition of $Cl_2$ to but-2-ene. (a) Write a mechanism for the reaction. (b) Is it likely that both *syn-* and *anti*-addition may occur? (c) How many stereoisomers are there of the product?
3. Outline the mechanism for the addition of $Br_2$ to but-2-ene, and suggest how many stereoisomers are likely to be formed.

## ADDITION POLYMERS

**Problem set 8.12**

1. Low-density polyethene may be produced by reaction 8.1. (a) What role does the dioxygen play? (b) During the polymerization reaction represented

in equation 8.1, some branching of the polymer chain occurs. Suggest how a methyl branch (substituent) may be introduced into the chain during the reaction. (c) Will branched hydrocarbon polymer chains pack more or less efficiently than linear chains in the solid state? (d) Why do you think that the product formed from reaction 8.1 is called *low-density* polyethene?

$$n \ H_2C{=}CH_2 \ \xrightarrow[\quad O_2 \quad]{470 \ K; \quad 1500 \ bar} \ \left( \begin{array}{cc} C & C \\ | & | \\ H_2 & H_2 \end{array} \right)_n \qquad (8.1)$$

2.    High-density polyethene is manufactured from ethene using a Ziegler-Natta catalyst. Keeping in mind your answers to question 1, what structural property of this polymer results in it possessing a relatively high density in the solid state?

**Ziegler-Natta catalyst: see Section 13.4 in H&C**

3.    Will high-density polyethene possess a higher or lower melting point than the low-density form? Rationalize your answer.

4.    A Ziegler-Natta catalyst is used in the manufacture of polypropene to give an *isotactic* polymer. (a) What does isotactic mean? (b) Why is it important to ensure that the addition of the propene monomers takes place in a controlled manner?

## REACTIONS OF ALKYNES

**Problem set 8.13**

1.    Determine the standard enthalpy of combustion of propyne if $\Delta_f H^\circ(H_2O, l)$ = –286; $\Delta_f H^\circ(CO_2, g)$ = –393.5; $\Delta_f H^\circ(C_3H_4, g)$ = 185 kJ mol$^{-1}$.

2.    Determine the standard enthalpy change for the hydrogenation of (a) propyne to propene, and (b) propene to propane. Comment on the relative magnitudes of your answers. [$\Delta_f H^\circ(C_3H_4, g)$ = 185; $\Delta_f H^\circ(C_3H_6, g)$ = 20; $\Delta_f H^\circ(C_3H_8, g)$ = –105 kJ mol$^{-1}$].

3.    When an excess of HBr adds to but-1-yne, the predominant product is 2,2-dibromobutane. Rationalize this observation.

4.    What product would you expect to obtain from the following reactions in which the compound being added to the alkyne is *in excess*?
(a) $CH_3C{\equiv}CH + Br_2 \rightarrow$
(b) $CH_3CH_2CH_2C{\equiv}CCH_3 + H_2 \xrightarrow{\text{Ni catalyst}}$
(c) $HC{\equiv}CCMe_2CH_3 + HCl \rightarrow$

5.    In the controlled two-step reaction of pent-1-yne with HBr followed by HI, what products will be produced both in the intermediate and final stages of the reaction?

6.    Amongst simple alkynes, why do only terminal ones function as weak acids?

7.    What products are obtained from the following reactions?
(a) $CH_3CH_2C{\equiv}CH + Na \rightarrow$
(b) $CH_3C{\equiv}CH + NaNH_2 \rightarrow$
(c) $CH_3C{\equiv}CCH_3 + KNH_2 \rightarrow$

8.    Water is a stronger acid than acetylene. How will water react with sodium acetylide?

9.    Acetylene may be prepared by the reaction of calcium carbide with water; one mole of calcium carbide is converted to one mole of acetylene. Calcium carbide is an ionic compound which adopts a distorted sodium chloride lattice.

(a) What is the formula of calcium carbide?

(b) Write an equation for the reaction of calcium carbide with water.

(c) Suggest how the ions in calcium carbide are arranged in the solid state.

## AN ADVANCED STUDY SECTION: OPTICAL ACTIVITY AND CHIRALITY

This section deals in more detail with chirality and refers to several terms that are defined in Chapter 14 of *Chemistry: An Integrated Approach*. You may wish to study this section now, or return to it after studying the material in Chapter 14.

In the main text book and in this workbook, we have discussed the property of chiralty and related it to molecular asymmetry and/or the presence of an asymmetric atomic centre within a molecule. Thus far, we have not dwelt upon the experimental methods for the observation of chiral compounds.

Pairs of enantiomers possess identical chemical properties — they only differ in their interaction with other chiral moieties. A familiar example of this is seen in the interaction of left- and right-hands with left and right gloves. The left-hand / left glove interaction is different from the left-hand / right glove interaction; these are in effect a pair of *diastereomers*.

➤ **Diastereomers: see Box 14.3 in H&C**

One of the simplest ways of distinguishing between enantiomers involves their interaction with other chiral systems. A very convenient method involves the interaction with polarized light. *Polarized light* is a form of light in which the electromagnetic wave lies in a single plane. When polarized light interacts with a chiral molecule, the plane in which the electromagnetic wave lies changes; in

**8.7**   A transverse wave of polarized light lies in one plane, in this case the plane of the paper.

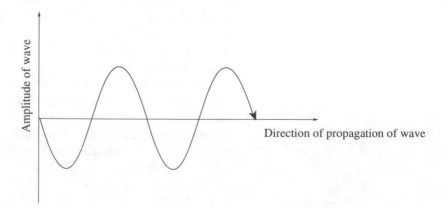

effect, this plane rotates around the axis of propagation of the light (Figure 8.7). A convenient way of characterizing a chiral compound is by measuring the amount of rotation of this plane. The rotation, $\alpha$, may be measured in a polarimeter. In practice, the amount of rotation depends upon the wavelength of the light, temperature and the amount of compound present in solution. The *specific rotation,* $[\alpha]$, for a chiral compound in solution is defined in equation 8.2.

$$[\alpha] = \frac{\alpha}{c \times \ell} \qquad (8.2)$$

where:    $\alpha$ = observed rotation
$\ell$ = path length of solution in the polarimeter (in dm)
$c$ = concentration (in g cm$^{-3}$)

➤
**Atomic emission spectra:
see page 21 of the
workbook**

It is very convenient to use light of a common frequency for such measurements and very often one of the lines from the atomic emission spectrum of sodium is used. This emission is known as the sodium D line, and the specific rotation at this wavelength is denoted as $[\alpha]_D$. The specific rotation of enantiomers is equal and opposite. For example, the two enantiomers of glyceraldehyde, structure **8.7**, have $[\alpha]_D$ values of $+11°$ and $-11°$. This leads to a common way of distinguishing between enantiomers by denoting the sign of $[\alpha]_D$ — the two enantiomers of glyceraldehyde are denoted (+)-glyceraldehyde and (–)-glyceraldehyde. Sometimes (+) and (–) are denoted by the prefixes *dextro-* and *laevo-* (derived from the Latin words for right and left) and indicate a right-handed and left-handed rotation of the light respectively. These are often abbreviated to *d* and *l*. We must emphasize here that the +/– or *d/l*-notations refer to the observed rotation of the light; as such they are temperature, concentration and wavelength dependent. They give no information, *a priori*, about the absolute configuration of the chiral compound.

(8.7)

Exactly the same ideas apply to the *chemical* separation of pairs of enantiomers; it is necessary to form *diastereomers* by reaction with a second chiral compound. The classical method of separating enantiomers involves conversion of the compound to a cation or anion, and forming a salt with a chiral counter-ion. Consider a chiral compound **XA** which can form a pair of enantiomeric cations (+)-**A**$^+$ and (–)-**A**$^+$. When treated with a chiral anion (–)-**B**$^-$, a pair of diastereomeric salts is formed (+)-**A**(–)-**B** and (–)-**A**(–)-**B**. Whereas enantiomers possess identical physical properties because the relative arrangements of atoms in space are the same, diastereomers possess different physical properties because the relative arrangements of atoms in space are different. Very often, pairs of diastereomers differ in their solubilities and the addition of (–)-**B**$^-$ to a racemic mixture of (+)-**A**$^+$ and (–)-**A**$^+$ results in the precipitation of only one of the diastereomeric salts (+)-**A**(–)-**B** or (–)-**A**(–)-**B**. This process of separation of enantiomers is known as *resolution* — the mixture is *resolved* into its component enantiomers.

➤
**Racemic mixture:
see Section 14.5 in H&C**

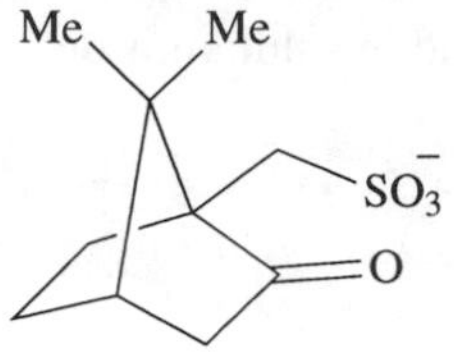

(8.8)

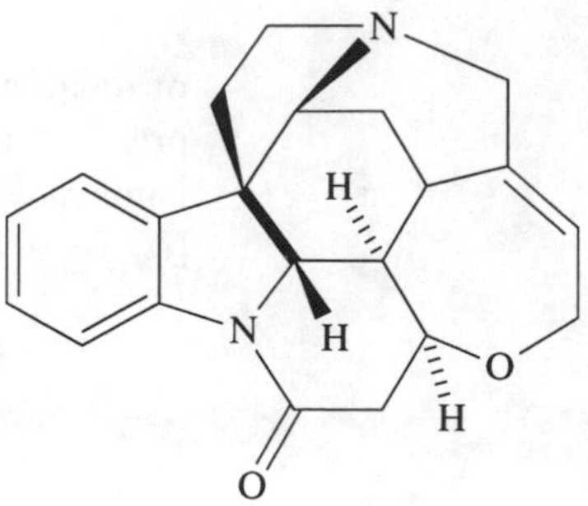

(8.9)

➤
**Alkaloids: see Box 15.10
in H&C**

Typical reagents which are used for the resolution of chiral compounds include anions such as **8.8**, or cations derived from *alkaloids* such as strychnine, structure **8.9**.

More modern methods of separation involve column chromatography (see page 61 of the workbook) using either a chiral solid phase in the column or a chiral eluting agent.

# 9 Spectroscopy

---

### Topics
- The Beer-Lambert Law
- IR spectroscopy in the laboratory
- Electronic spectroscopy in the laboratory
- NMR spectroscopy in the laboratory

---

## THE BEER-LAMBERT LAW

The Beer-Lambert Law (equation 9.1) relates the absorbance ($A$) of a solution to the molar extinction coefficient ($\varepsilon$, in $dm^3\,mol^{-1}\,cm^{-1}$), the concentration ($c$, in $mol\,dm^{-3}$), and the path length ($\ell$ in cm).

$$A = \varepsilon \times c \times \ell \tag{9.1}$$

**Worked example 9.1**

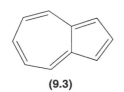

**(9.1)**

**(9.2)**

**Azobenzene (the IUPAC name for which is diphenyldiazene) possesses two geometrical isomers, 9.1 and 9.2, although 9.1 rapidly isomerizes to 9.2. Solutions of compound 9.1 absorb light of wavelength 247 nm and the extinction coefficient for this absorption is 11 500 $dm^3\,mol^{-1}\,cm^{-1}$. Solutions of compound 9.2 absorb at 316 nm with a corresponding extinction coefficient of 22 000 $dm^3\,mol^{-1}\,cm^{-1}$. Determine the relative absorbances of a $5 \times 10^{-3}$ M solution of 9.1 and a $2 \times 10^{-3}$ M solution of 9.2, each contained in a cell of 1 cm path length. [See also question 6 in problem set 9.1.]**

From equation 9.1, we can write:

$$\frac{A_1}{A_2} = \frac{\varepsilon_1 \times c_1 \times \ell_1}{\varepsilon_2 \times c_2 \times \ell_2}$$

As the path length is constant, $\ell_1$ and $\ell_2$ cancel. After substitution of the values given, we have:

$$\frac{A_1}{A_2} = \frac{11\ 500 \times 5 \times 10^{-3}}{22\ 000 \times 2 \times 10^{-3}}$$

The relative absorbances of the two solutions = $\dfrac{A_1}{A_2}$ = 1.3

---

**Problem set 9.1**

**(9.3)**

1.  Solutions of azulene **9.3** in cyclohexane absorb at 357 nm and the value of $\varepsilon$ for this absorption is 3980 $dm^3\,mol^{-1}\,cm^{-1}$. Such a solution contained in a cell of path length 1 cm gives an absorbance of 3.58. Determine the concentration of the solution.

2.  Solutions of phenanthrene **9.4** in cyclohexane absorb at 357 nm; this is one of four absorptions. In a series of experiments, the absorbance is measured as a function of solution concentration and the data obtained are

**(9.4)**

**(9.5)**

**(9.6)**

**(9.7)**

tabulated below:

| Concentration / mol dm$^{-3}$ | 0.0008 | 0.0012 | 0.0017 | 0.0030 | 0.0035 | 0.0050 |
|---|---|---|---|---|---|---|
| Absorbance | 0.167 | 0.220 | 0.355 | 0.627 | 0.720 | 1.045 |

(a) Use the data to determine the extinction coefficient for the absorption at 357 nm. (b) Why is it more accurate to measure $\varepsilon$ from a *series* of readings rather than a single reading in a cell of known path length?

3.  Solutions of acridine **9.5** in ethanol absorb at $\lambda = 250$ nm and for this absorption, $\varepsilon = 199\,500$ dm$^3$ mol$^{-1}$ cm$^{-1}$. (a) Determine the concentration of acridine in a solution within a cell of path length 0.5 cm for which the absorbance is 95.8. (b) What mass of acridine would be required to prepare a 250 cm$^3$ solution of this concentration?

4.  Solutions of naphthalene **9.6** in ethanol absorb at $\lambda = 312$ nm. A 0.250 M solution gives an absorbance of 72. Determine the concentration of a solution for which the absorbance is 100. The same solution cell is used for the two readings.

5.  The compound K$_3$[Fe(CN)$_6$] (the anion in which is shown in structure **9.7**) absorbs light of wavelength 418 nm ($\varepsilon = 1012$ dm$^3$ mol$^{-1}$ cm$^{-1}$). What mass of solid is needed to prepare a 200 cm$^3$ solution such that a sample of the solution, when contained in a cell of path length 1 cm, exhibits an absorbance of 0.61? [$A_r$ Fe = 56; C = 12; N = 14; K = 39]

6.  The conversion of compound **9.8** to its (Z)-isomer is induced by a flash of light (*flash photolysis*). Isomer **9.8** absorbs light of wavelength 435 nm whereas the (Z)-isomer absorbs at 316 nm. Figure 9.1 shows the change in absorbance at $\lambda = 435$ nm for a solution of **9.8** in cyclohexane over a period

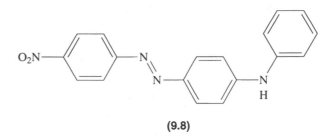

**(9.8)**

**9.1** The change in absorbance at $\lambda = 435$ nm for a cyclohexane solution of compound **9.8** before and after the solution was subjected to flash photolysis.

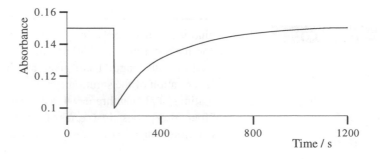

of 1200 s during which time the solution was subjected to a flash of light. (a) Rationalize the shape of the curve in Figure 9.1. (b) If the path length of the solution cell used in the experiment was 1 cm, and the solution concentration was $9 \times 10^{-6}$ mol dm$^{-3}$, determine a value for $\varepsilon$ corresponding to the absorption at 435 nm for **9.8**. [Data from: S.R. Hair, G.A. Taylor and L.W. Schultz, *Journal of Chemical Education* (1990) vol. 67, p. 709.]

## IR SPECTROSCOPY IN THE LABORATORY

➤ **Fundamental frequencies, force constants and reduced mass**

**Vibrations of diatomic molecules: see Section 9.6 in H&C**

**See also Worked Example 9.3 (page 408) in H&C**

The reduced mass, $\mu$, of a diatomic molecule may be determined using equation 9.2. Equation 9.3 relates the frequency measured in wavenumbers, $\bar{v}$, of the fundamental absorption in the IR spectrum of a diatomic molecule to the force constant, $k$, of the bond and the reduced mass, $\mu$. Equations 9.2 and 9.3 are also applicable to a pair of bonded atoms (e.g. a C–H or O–H unit) in a polyatomic molecule.

$$\frac{1}{\mu} = \frac{1}{m_1} + \frac{1}{m_2} \tag{9.2}$$

where $m_1$ and $m_2$ are the masses in kg of the two atoms

$$\bar{v} = \frac{1}{2\pi \times c} \sqrt{\frac{k}{\mu}} \qquad \text{where } c = \text{speed of light} \tag{9.3}$$

**Problem set 9.2**

**Data needed for this problem set:**
Speed of light,
    $c = 3.0 \times 10^8$ m s$^{-1}$
Atomic mass unit = $1.66 \times 10^{-27}$ kg
$A_r$ H = 1; C = 12; N = 14; O = 16; Br = 80; S = 32

1. Determine the reduced masses of (a) a CO molecule, (b) a C–H unit, (c) an S–H unit, and (d) an HD molecule (D = deuterium = $^2$H).

2. Determine the fundamental stretching frequency for HBr if the force constant for the H–Br bond is 412 N m$^{-1}$.

3. Determine the force constant for the C≡N bond in HCN if the fundamental stretching frequency is 2220 cm$^{-1}$.

4. The force constants for the bonds in the molecules ClF, BrF and BrCl are 448, 406 and 282 N m$^{-1}$ respectively. Predict the *trend* in the bond dissociation enthalpies of these bonds.

5. If the force constant for an S–H bond is 428 N m$^{-1}$, determine the fundamental stretching frequency for this bond.

6. *Deuteration* is the process of exchanging $^1$H nuclei in a molecule for $^2$H (i.e. D) nuclei; for example, D$_2$CO **9.9** is deuterated methanal. Methanal, H$_2$CO, exhibits an absorption due to the fundamental C–H stretch at 2783 cm$^{-1}$. What will be the *shift* in the fundamental frequency for the C–X (X = H or D) bonds upon deuterating methanal? Will the shift be to higher or lower wavenumber? [Assume that the force constants $k$(C–H) $\approx$ $k$(C–D).]

**(9.9)**

**9.2** Part of the infrared spectrum of the compound $CH_3CH_2CH_2C{\equiv}N$ showing the absorption for the stretching mode of the $C{\equiv}N$ bond.

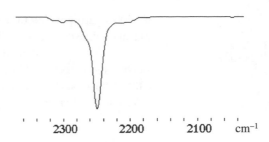

7.    Figure 9.2 shows the absorption in the IR spectrum of $CH_3CH_2CH_2C{\equiv}N$ due to the stretch of the $C{\equiv}N$ bond. If this compound were isotopically labelled with (a) $^{13}C$ or (b) $^{15}N$, would you expect the absorption shown in the figure to show a significant shift in wavenumber? Rationalize your answer. [Natural isotopic abundances: $^{12}C$ 98.9%; $^{14}N$ 99.6%.]

➤ **An introduction to the interpretation of infrared spectra**

**Further practice in interpreting IR spectra is given in Chapters 14, 15 and 17 of this workbook**

The following problems aim to provide you with practice in interpreting IR spectra in the laboratory. You should be able to pick out characteristic absorptions due to certain functional groups, and also be able to recognize absorptions which come in the *fingerprint region*. You may need to refer to Tables 9.3 to 9.6 in H&C; these tables list some of the characteristic absorptions exhibited by functional groups.

**Problem set 9.3**

1.    Figure 9.3 shows the IR spectrum of ethanol. (a) What is the structure of ethanol. (b) In the IR spectrum, to what do you assign the band centred at $3340$ cm$^{-1}$? (c) To what do you assign the group of absorptions between $2890$ and $2970$ cm$^{-1}$? (d) In what ways would you expect the IR spectrum of propan-1-ol to be similar to or different from that of ethanol?

2.    Why would you not expect to observe an infrared spectroscopic absorption due to the stretch of the $C{\equiv}C$ bond in acetylene? Would you expect to observed such an absorption in the IR spectrum of $CH_3C{\equiv}CH$? Rationalize your answers.

3.    The IR spectrum of hydrogen cyanide, HCN, shows absorptions at 3311, 712 and 2097 cm$^{-1}$. Suggest assignments for these bands.

**9.3** The IR spectrum of ethanol.

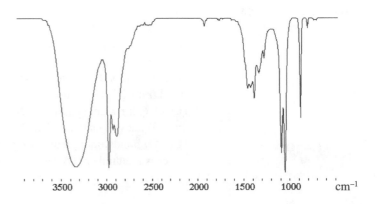

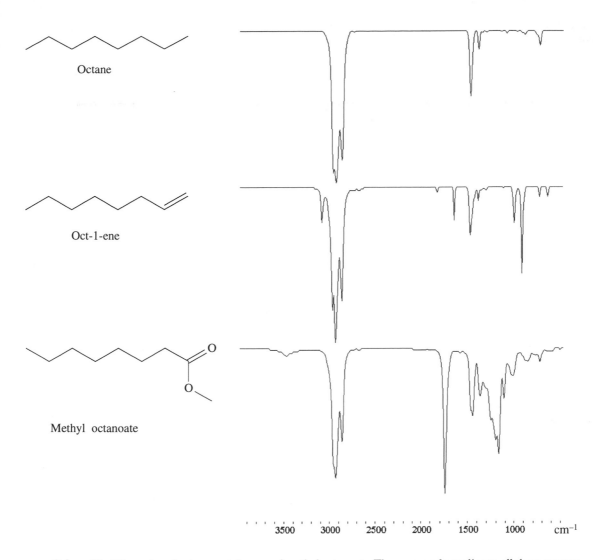

**9.4**    The IR spectra of octane, oct-1-ene and methyl octanoate. The same scale applies to all three spectra.

4.    Figure 9.4 shows the IR spectra of octane, oct-1-ene and methyl octanoate. Which absorptions in these spectra can you readily assign?

5.    The IR spectrum of $CO_2$ exhibits absorptions at 2349 and 667 cm$^{-1}$, while that of $SO_2$ shows bands at 1362, 1151 and 518 cm$^{-1}$. (a) Draw the structures of $CO_2$ and $SO_2$. (b) Why does the IR spectrum of $CO_2$ exhibit two absorptions, whilst that of $SO_2$ shows three? (c) To what mode of vibration do you assign the absorption at 2349 cm$^{-1}$ in the spectrum of carbon dioxide?

6.    When oct-1-ene (Figure 9.4) reacts with ozone followed by reductive hydrolysis, two products are obtained, both of which exhibit strong absorptions around 1750 cm$^{-1}$ in their IR spectra. Rationalize these data.

**9.5**  The IR spectrum of
2-chloro-2-methylpropane.

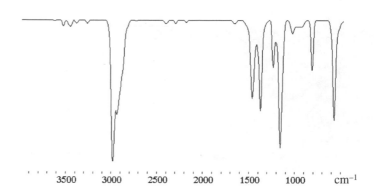

7.   Figure 9.5 shows the IR spectrum of 2-chloro-2-methylpropane. (a) Draw the structure of 2-chloro-2-methylpropane. (b) Is it possible to assign unambiguously the absorption due to the C–Cl stretching mode?

8.   (a) What shape is an $SO_3$ molecule? (b) Draw a diagram to show the symmetric stretching mode of vibration in $SO_3$. (c) Is this mode of vibration IR active?

9.   Figure 9.6 shows the IR spectrum of a chlorinated derivative of buta-1,3-diene. What information about the compound can you deduce from the spectrum?

10.  Figure 9.7 shows parts of two spectra labelled **A** and **B**, and two compounds, **I** and **II**. Assign the two spectra to the two compounds, giving reasons for your choices.

11.  Figure 9.8 shows the structure and infrared spectrum of the drug cocaine. (a) Assign as many spectroscopic bands as you are able. (b) How useful would IR spectroscopy be as a method for customs officers to verify the seizure of a batch of cocaine? For example, could this technique distinguish between cocaine and aspirin, structure **9.10**?

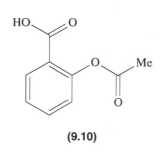

(9.10)

**9.6**  IR spectrum for
question 9, problem set 9.3.

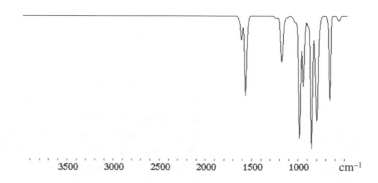

**9.7** Infrared spectra for problem 10, problem set 9.3.

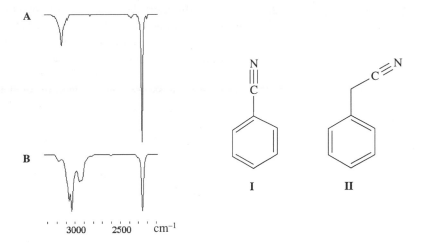

**9.8** The structure and IR spectrum of cocaine; for question 11, problem set 9.3.

## ELECTRONIC SPECTROSCOPY IN THE LABORATORY

### Fundamental concepts of electronic spectroscopy

Electronic spectroscopy involves transitions between electronic energy levels. The following set of problems looks at some basic features of electronic transitions and the observation of electronic spectra.

**Problem set 9.4**

1.  Electronic spectra are often reported in terms of the wavelength of the absorbed electromagnetic radiation. Convert the following absorption maxima into wavelengths given in nm: (a) $3 \times 10^{-8}$ m; (b) $3.5 \times 10^{-9}$ cm; (c) 22 700 cm$^{-1}$.

2.  Normal laboratory UV–VIS spectrometers operate in the approximate wavelength range of 200 to 900 nm. Which of the following types of electromagnetic radiation does this range cover: (a) visible light; (b) near-UV radiation, (c) vacuum-UV radiation?

3.  Do absorptions in the vacuum-UV involve transitions in which $\Delta E$ (change in energy) is greater or smaller than those observed in the visible region?

4.  (a) For what do the abbreviations HOMO and LUMO stand? (b) A related abbreviation is NBMO. For what do you think this stands?

5.  (a) Why are molecular electronic spectra usually characterized by *broad* absorptions? (b) Electronic absorptions are usually denoted in terms of $\lambda_{max}$ and $\varepsilon_{max}$. Draw a diagram of a typical absorption in an electronic spectrum and indicate how you would measure $\lambda_{max}$ and $\varepsilon_{max}$. (c) Why do we specify $\lambda_{max}$ and $\varepsilon_{max}$, and not simply $\lambda$ and $\varepsilon$ ?

6.  Table 9.1 lists values of $\lambda_{max}$ for some inorganic and organic species. In what part of the electromagnetic spectrum does each absorb?

7.  Table 9.2 lists absorbance data for three solvents which might be used to prepare samples for measurements of their electronic spectra. For each solvent, the increase in absorbance at lower wavelength indicates the onset of an absorption band. (a) If you wished to record the UV-VIS spectrum of benzene ($\lambda_{max}$ = 183, 204 and 256 nm), would cyclohexane be a suitable solvent? Explain your answer. (b) Which of the solvents listed are polar? How does this property affect the ability of each to act as a solvent? (c) Pyrazine **9.11** is soluble in both cyclohexane and acetone. Which solvent would you choose to use when recording the UV-VIS spectrum of pyrazine ($\lambda_{max}$ = 260 and 327 nm)? (d) Which of the solvents listed would be suitable for use in recording the electronic spectra of compounds that absorb in the visible region?

**(9.11)**

**Table 9.1**    Absorption maxima for selected species.

| Species | Formula | $\lambda_{max}$ / nm |
|---|---|---|
| Iodide ion | $I^-$ | 226 |
| Oxalate ion | $[C_2O_4]^{2-}$ | 250 |
| Azide ion | $[N_3]^-$ | 235 |
| Water | $H_2O$ | 167 |
| Acetylene | $HC \equiv CH$ | 173 |
| Aqueous Ni(II) ions | $[Ni(H_2O)_6]^{2+}$ | 440[a] |
| Azobenzene[b] [(Z)-isomer] | $C_6H_5N=NC_6H_5$ | 316 |

[a]This is one of three absorptions.
[b]See structure **9.2** on page 75.

**Table 9.2**  Electronic spectroscopic data for some common solvents

**Dimethylformamide (DMF)**

| Wavelength | 268 | 275 | 300 | 350 | 400 |
|---|---|---|---|---|---|
| Max. absorbance | 1.000 | 0.300 | 0.050 | 0.005 | 0.005 |

**Cyclohexane**

| Wavelength | 200 | 225 | 250 | 300 | 400 |
|---|---|---|---|---|---|
| Max. absorbance | 1.000 | 0.170 | 0.020 | 0.005 | 0.005 |

**Acetone**

| Wavelength | 330 | 340 | 350 | 375 | 400 |
|---|---|---|---|---|---|
| Max. absorbance | 1.000 | 0.060 | 0.010 | 0.005 | 0.005 |

## $\pi$-Conjugation

**Problem set 9.5**

1. What is a *polyene*?
2. In which of the following compounds is the $\pi$-bonding delocalized? (a) buta-1,3-diene; (b) penta-1,4-diene; (c) hexa-1,5-diene; (d) hexa-1,3,5-triene.
3. Figure 9.9 shows the dependence of $\lambda_{max}$ on the number of C=C double bonds in members of the series of polyenes $C_6H_5(HC=CH)_nC_6H_5$. (a) Why does the absorption maximum shift to longer wavelength? (b) In what part of the spectrum does each compound absorb (e.g. near-UV, visible)?

➤ **The structure of lycopene is given in Box 9.3 in H&C**

4. Lycopene ($\lambda_{max} = 469$ nm) is present in tomatoes. What colour of light does lycopene absorb?
5. (a) What is a *red shift*? (b) Explain why the methoxy substituent in the compound $MeOCH=CHCH=CH_2$ causes a red shift in the UV-VIS spectrum with respect to the spectrum of $CH_2=CHCH=CH_2$.

**9.9**  Values of $\lambda_{max}$ for the $\pi \rightarrow \pi^*$ transitions in the polyenes $C_6H_5(HC=CH)_nC_6H_5$ for values of $n = 3$–6.

$C_6H_5$ = phenyl group; see Chapter 15.

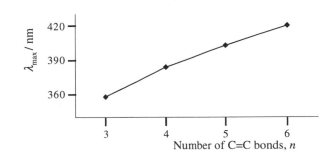

## Applications of UV-VIS spectroscopy

**Problem set 9.6**

1.  The reaction shown in equation 9.4 can be followed by UV-VIS spectroscopy. With the spectrometer tuned to a wavelength of 470 nm, the results shown in Figure 9.10 were obtained. What can you deduce about the absorbance of the components in equation 9.4?

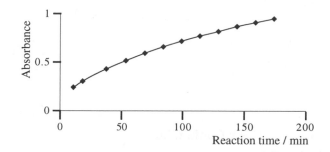

$$2 \ O_2N-\underset{}{\bigcirc}-CH_2Cl \xrightarrow[- \ 2HCl]{2[CN]^-} O_2N-\bigcirc-CH=CH-\bigcirc-NO_2 \quad (9.4)$$

**9.10** Absorbance at 470 nm measured as a function of time during the reaction shown in equation 9.4. [Data from: M.O. Hurst and J.W. Hill, *Journal of Chemical Education* (1993) vol. 70, p. 429.]

2.  The reaction between $I^-$ and $[S_2O_8]^{2-}$ ions occurs according to equation 9.5 but the $I_2$ produced is present in the form of $[I_3]^-$ ions so long as an excess of $I^-$ is present (equation 9.6). For $[I_3]^-$, $\lambda_{max} = 353$ nm, $\varepsilon_{max} = 26\ 000$ dm$^3$ mol$^{-1}$ cm$^{-1}$.

$$[S_2O_8]^{2-} + 2I^- \rightarrow 2[SO_4]^{2-} + I_2 \quad (9.5)$$
$$I_2 + I^- \rightarrow [I_3]^- \quad (9.6)$$

    Explain how measurements of the absorbance at $\lambda = 353$ nm would provide you with a measure of the change in concentration of $I_2$ during the reaction.

3.  A ligand $L^{4-}$ reacts with aqueous chromium(III) ions to give a coloured complex $[CrL_x]^{n-}$. In order to determine the stoichiometry of this complex anion, 10 cm$^3$ solutions composed of $V$ cm$^3$ of a 0.002 M aqueous chromium(III) nitrate and $(10 - V)$ cm$^3$ 0.002 M aqueous $H_4L$ were prepared. With a colorimeter tuned to absorb light of a wavelength corresponding to $\lambda_{max}$ for $[CrL_x]^{n-}$, the variation in absorbance as a function of solution composition (a Job's plot) was recorded. Using the following results, determine the values of $x$ and $n$ in $[CrL_x]^{n-}$.

| $V$ / cm$^3$ | 0 | 1.0 | 2.0 | 3.0 | 4.0 | 6.0 | 7.0 | 8.0 | 9.0 | 10.0 |
|---|---|---|---|---|---|---|---|---|---|---|
| Absorbance | 0 | 0.21 | 0.40 | 0.59 | 0.81 | 0.80 | 0.60 | 0.38 | 0.20 | 0 |

## NMR SPECTROSCOPY IN THE LABORATORY

➤
**Further practice in interpreting NMR spectra is given in Chapters 14, 15 and 17 of this workbook**

The problem sets dealing with NMR spectroscopy are divided into three groups. The first deals with $^{13}C$ NMR spectroscopy and focuses on:
- identifying the number of different types of carbon centres in a molecule;
- identifying particular types of carbon atom such as $sp^2$ and $sp^3$ centres.

The second section deals with $^{1}H$ NMR spectroscopy and includes practice in the interpretation of simple $^{1}H–^{1}H$ spin-spin coupling patterns. For this you may find it helpful to refer to Pascal's triangle in Figure 9.39 in H&C.

The final section contains more advanced problems and includes examples of heteronuclear coupling.

### $^{13}C$ NMR spectroscopy

**Problem set 9.7**

**Refer to Table 9.1 in H&C**

1. The $^{13}C$ NMR spectrum of a monoiodo-derivative of propane exhibits three signals of equal intensity at $\delta$ 15.3, 26.8 and 9.2. Which isomer of iodopropane is this?

2. (a) Draw the structures of the isomers of $C_4H_9Br$. (b) The shifts and relative integrals of the $^{13}C$ NMR spectroscopic signals of three of the isomers (labelled **I**, **II** and **III**) are as follows:

| Isomer **I** | $\delta$ | 42.4 | 30.7 | 20.9 | | [1 : 1 : 2] |
|---|---|---|---|---|---|---|
| Isomer **II** | $\delta$ | 53.1 | 34.2 | 26.0 | 12.1 | [1 : 1 : 1 : 1] |
| Isomer **III** | $\delta$ | 62.1 | 36.4 | | | [1 : 3] |

Assign each spectrum to a particular isomer. Are the assignments unambiguous?

3. The reaction between $Cl_2$ and ethene leads to a product with a $^{13}C$ NMR spectroscopic signal at $\delta$ 51.7, whilst the addition of one equivalent of $Cl_2$ to ethyne leads to two products; one product has a signal at $\delta$ 119.0 in its $^{13}C$ NMR spectrum and the other at $\delta$ 121.0. Account for these observations.

4. Figure 9.11 shows the $^{13}C$ NMR spectrum of $N\equiv CCH_2CH_2CH_2CH_2C\equiv N$. (a) What is the hybridization of each carbon centre? (b) Which signal(s) in the spectrum can you unambiguously assign?

**9.11** 25 MHz $^{13}C$ NMR spectrum of $NC(CH_2)_4CN$.

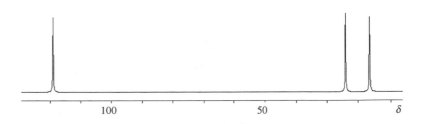

**9.12**  25 MHz $^{13}$C NMR
spectra of
(a) (E)-HBrC=CHMe and
(b) (Z)-HBrC=CHMe.

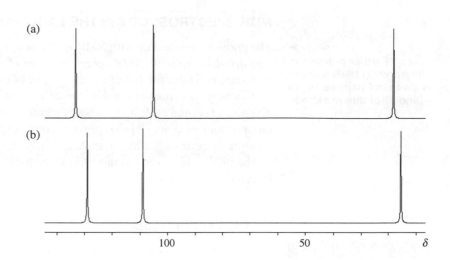

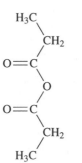

(9.12)

5.  The acid-catalysed addition of water to 2-methylpropene gives a product
    with an $^{13}$C NMR spectrum consisting of two signals at $\delta$ 31.2 and 68.9
    with relative integrals of 3:1, respectively. Explain how these data confirm
    that Markovnikov addition has occurred.

6.  (a) Draw the structures of (E)-HBrC=CHMe and (Z)-HBrC=CHMe.
    (b) Figure 9.12 shows the $^{13}$C NMR spectra of these isomers. How useful
    would $^{13}$C NMR spectroscopy be in identifying which isomer you had
    produced on the addition of HBr to propyne?

7.  An isomer of tetrachloroethane exhibits one signal in its $^{13}$C NMR spectrum
    ($\delta$ 6.0). Give the name of this isomer.

8.  The $^{13}$C NMR spectrum of the anhydride **9.12** has three signals at $\delta$ 8.4,
    28.7 and 170.3. Assign the spectrum.

9.  Figure 9.13 gives the $^{13}$C NMR spectrum of neat DMF which is a common
    laboratory solvent. The molecular structure of DMF was given on page 83.
    Assign the spectrum and explain why there are three, rather than two, signals.

**9.13**  25 MHz $^{13}$C NMR
spectrum of DMF.

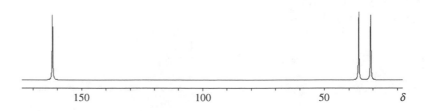

## ¹H NMR spectroscopy

The problems in this section require that you have studied *both* the diagnostic shifts and ¹H-¹H coupling patterns that are observed in ¹H NMR spectra (see Sections 9.15 and 9.16 in H&C).

**Problem set 9.8**

**Refer to Table 9.12 in H&C**

$$CH_3C \overset{\displaystyle O}{\underset{\displaystyle OH}{\diagup\diagdown}}$$

**(9.13)**

$$HO\overset{\displaystyle Me}{\underset{\displaystyle Me}{-\overset{|}{\underset{|}{C}}-}}C\equiv C-H$$

**(9.14)**

1. The ¹H NMR spectrum of acetic acid **9.13** consists of two singlets at $\delta$ 2.1 and 11.4. (a) Assign the spectrum. (b) What other information from the spectrum would assist in the assignments?

2. The ¹H NMR spectrum of compound **9.14** consists of a broadened signal at $\delta$ 2.2, a singlet (relative integral 6) at $\delta$ 1.5 and a singlet (relative integral 1) at $\delta$ 2.4. Assign the spectrum.

3. Rationalize why the ¹H NMR spectrum of 2-iodopropane consists of a doublet and a septet.

4. Figure 9.14 shows the ¹H NMR spectrum of 1,1,2-trichloroethane. Rationalize the spectrum.

5. The ¹H NMR spectrum of bromoethane consists of signals at $\delta$ 1.7 and 3.4. Assign the signals and predict their coupling patterns.

6. The ¹H NMR spectrum of an isomer of trichloropropane consists of two singlets at $\delta$ 2.2 and 4.0 with relative integrals of 3 : 2 respectively. Identify the isomer present.

7. The ¹H NMR spectrum of propan-2-ol consists of signals at $\delta$ 1.2, 1.6 (broad) and 4.0. Assign the signals and predict their coupling patterns and relative integrals.

8. Figure 9.15 shows the ¹H NMR spectrum of 1-nitropropane, $CH_3CH_2CH_2NO_2$. (a) Account for the coupling patterns observed. (b) Would ¹H NMR spectroscopy be a useful method of distinguishing between 1-nitropropane and 2-nitropropane? Explain your answer.

**9.14**  100 MHz ¹H NMR spectrum of 1,1,2-trichloroethane.

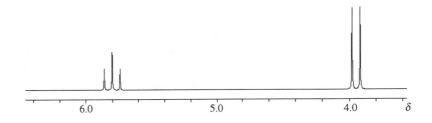

**9.15**  100 MHz ¹H NMR spectrum of 1-nitropropane.

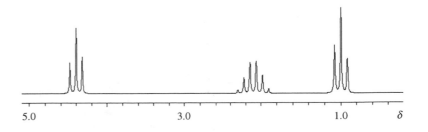

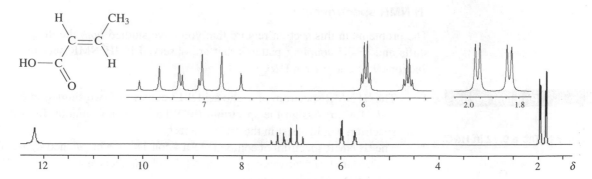

**9.16**  60 MHz ¹H NMR spectrum of (*E*)-but-2-enoic acid (crotonic acid), the structure of which is shown above. The figure shows the whole spectrum and also expansions of the multiplets at δ 1.90, 5.86 and 7.07. The scales of the expanded regions are *not* the same.

## Advanced study problems: More complex ¹H-¹H coupling and heteronuclear coupling

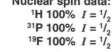

**Nuclear spin data:**
¹H 100% *I* = ¹/₂
³¹P 100% *I* = ¹/₂
¹⁹F 100% *I* = ¹/₂

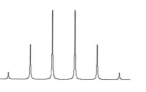

**9.17** The ³¹P NMR spectrum of PF₅ is a sextet.

1.  Figure 9.16 shows the ¹H NMR spectrum of an unsaturated carboxylic acid. (a) Which signal is assigned to the OH proton? (b) ¹H-¹H coupling constants of 1, 7.5 and 16 Hz can be measured from the spectrum. Mark on the spectrum where these values of $J_{HH}$ are measured. (c) Explain how the observed coupling patterns arise.

2.  How would you use ¹H NMR spectroscopy to distinguish between 1-chlorobutane, 2-chlorobutane and 2-chloro-2-methylpropane?

3.  (a) Use VSEPR theory to predict the structure of PF₅. How many fluorine environments are there? (b) The ³¹P NMR spectrum of PF₅ (recorded at 298 K) consists of a sextet (Figure 9.17). Explain how this arises. (c) Predict the nature of the signal in the room temperature ¹⁹F NMR spectrum of PF₅.

4.  (a) Predict the structure of the [PF₆]⁻ anion. (b) Predict the nature of the signals in the ¹⁹F and ³¹P NMR spectra of a solution of [NH₄][PF₆] (ammonium hexafluorophosphate).

5.  The ¹H NMR spectrum of CF₃CH₂OH consists of a quartet and a broad peak. How do you rationalize these observations?

6.  Figure 9.18 shows the ¹³C NMR spectrum of trifluoroacetic acid CF₃CO₂H. (a) Why are two quartets observed? (b) Assign the two signals.

**Problem set 9.9**

**9.18** The 25 MHz ¹³C NMR spectrum of CF₃CO₂H.

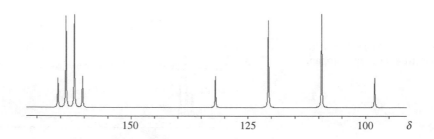

# 10 Reaction kinetics

> **Topics**
> - Growth and decay curves
> - Initial rates method
> - Data analysis
> - The Arrhenius equation
> - Elementary steps and the steady-state approximation

## GROWTH AND DECAY CURVES

A measure of the rate of a reaction is the change in concentration of a reactant or product as a function of time. For a reaction of stoichiometry:

$$- \ A \rightarrow B$$

the *decay curve* for A mirrors the growth of B as we showed in Figure 10.1 in H&C. We now consider growth and decay curves for other representative reaction types.

### 2A → B

In this case two moles of A are used to form one mole of B, and so the rate at which A decays must be twice that at which B is formed. Figure 10.1 illustrates the changes in concentrations of A and B for such a reaction.

**10.1** Growth of product B and decay of reactant A in a reaction of the general type:
$$2A \rightarrow B$$

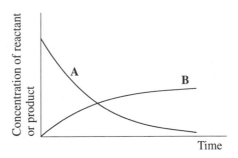

### Reactions with intermediates and more than one product

We saw in Chapter 8 that reactions may proceed via intermediate species and that more than one product may be formed. Figure 10.2 shows the results of a reaction in which reactant A decays to form two products, but during the reaction two

**10.2** The variation in the concentrations of reactant A and the intermediate and product species during the bromination of 1,2-dichlorobenzene. [Data from: D.A. Annis, D.M. Collard and L.A. Bottomly, *Journal of Chemical Education* (1995) vol. 72, p. 460.]

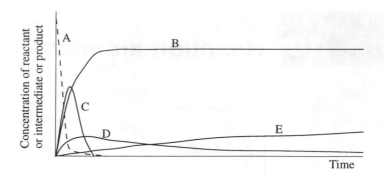

intermediate species are formed. *Exercise*: Which species are (a) intermediates, (b) the major product, and (c) the minor product?

## INITIAL RATES METHOD

One experimental strategy that we did not explore in Chapter 10 of the main text was the *initial rates method*. In this method, the rate is measured at the instant the reaction begins and the experiment is repeated several times using different concentrations of reactant. From these data, the dependence of the rate on the initial concentration of a reactant can be determined.

Consider a reaction of the type:        $A \rightarrow B$

The rate of the reaction is given by equation 10.1.

$$\text{Rate } (R) = k[A]^n \qquad \text{where } k = \text{rate constant,} \qquad (10.1)$$
$$n = \text{order of reaction with respect to A}$$

### Case 1: $n = 0$

➤ **Zero order reactions: see Section 10.2 in H&C**

For a zero order reaction, the plot of [A] against time is linear and the gradient is equal to the rate constant $k$ (Figure 10.3). It is not necessary to run more than one experiment since both the order and the value of $k$ can be confirmed from one set of data.

*Exercise*: How do you measure $k$ ?

**10.3** The change in [A] for a reaction of the type:
$$A \rightarrow B$$
which is zero order with respect to A.

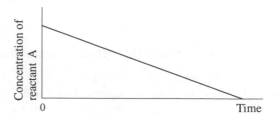

**10.4** The initial rates method. (a) For the reaction:
$$A \rightarrow B$$
the variation in [A] with time, $t$, is measured for *at least* three different values of $[A]_0$. The gradient of a tangent drawn to each curve at $t = 0$ is measured and this gives the *initial rate*, $R$, for each value of $[A]_0$.
(b) A graph is plotted of $\log R$ against $\log [A]_0$; this is a straight line from which the rate constant, $k$, and the order, $n$, with respect to A can be found (equation 10.3).

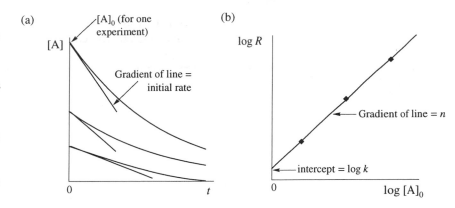

(a) [A] [A]$_0$ (for one experiment)    Gradient of line = initial rate    0    $t$

(b) $\log R$    Gradient of line = $n$    intercept = $\log k$    0    $\log [A]_0$

## Case 2: $n = 1$ or $2$

If a reaction shows first or second order kinetics with respect to A, a plot of [A] against time is a curve. By using different *initial concentrations* of A, $[A]_0$, different curves may be obtained (Figure 10.4a). The initial rate for each experiment (i.e. the rate of the reaction at $t = 0$) is found by measuring the gradient of a tangent drawn to each curve at $t = 0$. By comparing the values of the gradients, we can see whether, for example, the rate doubles when the concentration of A is doubled; this particular result would indicate that the rate depended directly on [A] and that the reaction was first order (equation 10.2).

$$\text{Rate} = k[A] \qquad \text{i.e.} \quad \text{Rate} \propto [A] \tag{10.2}$$

More generally we can treat the data as follows. By taking logarithms of the left- and right-hand sides of equation 10.1, we can write equation 10.3. Now compare equation 10.3 with equation 10.4, the equation for a straight line.

$$\log x^y = y \times \log x$$

$$\log (x \times y) = \log x + \log y$$

$$\log R = \log (k[A]^n) = \log k + (n \times \log [A]) \tag{10.3}$$

$$y = c + mx \quad \text{where } c = \text{intercept on the } y \text{ axis, and } m = \text{gradient} \tag{10.4}$$

Using the initial rate values from a series of experiments with different *initial* concentrations of A, plot a graph of $\log R$ against $\log [A]_0$ as is shown in Figure 10.4b. The rate constant $k$ is found from the intercept on the $\log R$ axis (at the point where $\log [A]_0 = 0$), and the order $n$ is equal to the gradient.

An advantage of the initial rates method is that the reaction need only be followed for a short period, and not until completion. The disadvantage of the method is that the measurement of the gradient of a tangent drawn to a curve (Figure 10.4a) is not an accurate means of determining a rate of reaction.

Questions 9 and 10 in problem set 10.1 give you practice in the initial rates method.

## DATA ANALYSIS

The problems in this section are designed to give you practice in analysing kinetic data. You should have studied worked examples 10.1 to 10.5 in H&C before attempting the questions.

**Problem set 10.1**

1.  Figure 10.5 shows the change in concentration of benzyl alcohol during the following reaction:

$$C_6H_5CH_2OH \ + \ CF_3CO_2H \ \longrightarrow \ CF_3CO_2CH_2C_6H_5 \ + \ H_2O$$

    benzyl alcohol    trifluoroacetic acid    benzyl trifluoroacetate    water

    Use the half-life method to confirm that the reaction is first order with respect to the alcohol.

**10.5** The change in concentration of benzyl alcohol during its reaction with trifluoroacetic acid. [Data from: D.E. Minter and M.C. Villareal, *Journal of Chemical Education* (1985) vol. 62, p. 911.]

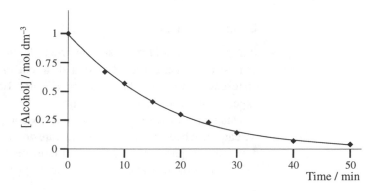

2.  Permanganate ion $[MnO_4]^-$ is an oxidizing agent and the kinetics of the following oxidation of an alcohol functionality have been studied:

$$[C_6H_5CH(OH)CO_2]^- \ \longrightarrow \ [C_6H_5C(O)CO_2]^-$$

➤

**Beer-Lambert Law: see equation 9.1 in the workbook**

    $[MnO_4]^-$ absorbs at 546 nm and the change in absorbance at this wavelength during the reaction can be used to monitor the change in concentration of $[MnO_4]^-$. If the concentration of $[C_6H_5CH(OH)CO_2]^-$ greatly exceeds the concentration of $[MnO_4]^-$, use the following data to determine the order of the reaction with respect to $[MnO_4]^-$.
    [Data from: R.D. Crouch, *Journal of Chemical Education* (1994) vol. 71, p. 597.]

| Time / min | 1.5 | 2.0 | 2.5 | 3.0 | 3.5 | 4.0 | 4.5 | 5.0 | 5.5 |
|---|---|---|---|---|---|---|---|---|---|
| Absorbance[a] | 0.081 | 0.072 | 0.062 | 0.054 | 0.047 | 0.040 | 0.035 | 0.031 | 0.027 |

[a]The absorbance data have been corrected for the fact that the absorbance does not reach zero when $[MnO_4]^- = 0$.

3. Every radioactive nuclide, N, decays with a characteristic half-life. Would a plot of ln [N] against time be linear or non-linear? Justify your answer.

4. The rate equation for a particular reaction of the type:

$$A + B \rightarrow products$$

is of the form:

$$Rate = k[A]^x[B]$$

If kinetic runs are carried out with A in vast excess with respect to B, the equation can be rewritten in the form:

$$Rate = k_{obs}[B]$$

**Hint**: look back at the margin note on page 91

(a) What is the name given to $k_{obs}$ ? (b) Use the following data to determine values of $x$ and $k$. (c) What is the overall order of the reaction?

| [A] / mol dm$^{-3}$ | 0.001 | 0.002 | 0.003 | 0.004 |
|---|---|---|---|---|
| $k_{obs}$ / min$^{-1}$ | 0.16 | 0.31 | 0.49 | 0.68 |

5. The reaction between permanganate ion and oxalate ion proceeds as follows:

$$2[MnO_4]^- + 5[C_2O_4]^{2-} + 16H^+ \longrightarrow 2Mn^{2+} + 10CO_2 + 8H_2O$$

Kinetic data for this reaction are tabulated below and show the results of carrying out an experiment in which the initial concentrations of $[C_2O_4]^{2-}$ and $H^+$ greatly exceeded those of $[MnO_4]^-$.

| Time / s | 330 | 345 | 360 | 375 | 390 | 405 | 420 |
|---|---|---|---|---|---|---|---|
| $[MnO_4^-] \times 10^3$ / mol dm$^{-3}$ | 1 | 0.487 | 0.223 | 0.067 | 0.030 | 0.011 | 0.005 |

**Note the way in which the data are given:**
$[MnO_4]^- = 1 \times 10^{-3}$ mol dm$^{-3}$
or
$[MnO_4]^- \times 10^3 = 1$ mol dm$^{-3}$

(a) Which species is the reducing agent in the reaction? (b) What was the purpose of having large initial concentrations of oxalate and hydrogen ions with respect to that of the permanganate ion? (c) Determine the order of the reaction with respect to $[MnO_4]^-$.

[Based on data from: B. Miles and S.K. Nyarku, *Journal of Chemical Education* (1990) vol. 67, p. 269.]

6. A compound X reacts to give a compound Y and Figure 10.6 shows the changes in concentrations of X and Y during the reaction. Write a balanced equation for the reaction, explaining how you arrive at your answer.

**10.6** For question 6, problem set 10.1; changes in the concentrations of X and Y during the reaction:

$$aX \rightarrow bY$$

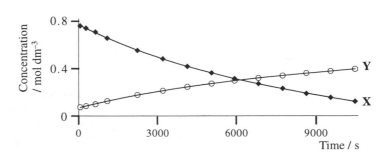

7.  The concentration of reactant A during a particular reaction of the type:
    $$A + 2B \rightarrow products$$
    changes as shown in the table below:

| Time / min | 1 | 5 | 10 | 15 | 25 | 40 |
|---|---|---|---|---|---|---|
| [A] / mol dm$^{-3}$ | 0.317 | 0.229 | 0.169 | 0.130 | 0.091 | 0.062 |

The reaction is $n$ order with respect to A and zero order with respect to B.
(a) Show that $n = 2$. (b) Write a rate equation for the reaction. (c) Determine the rate constant.

**Diels-Alder reaction: see Section 15.4 in H&C**

8.  In a Diels-Alder reaction, a double bond adds across the 1,4-positions of a 1,3-conjugated diene. An example is shown below:

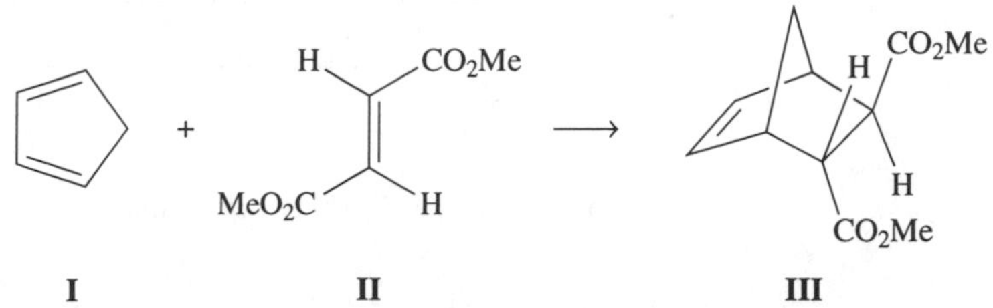

**I**                **II**                **III**

A study of the kinetics of this reaction, in which the initial concentrations of compounds **I** and **II** were *equal,* have yielded the following results:

| Time / s | 0 | 240 | 840 | 1800 | 2880 | 4200 | 5520 | 6000 |
|---|---|---|---|---|---|---|---|---|
| [**II**] / mol dm$^{-3}$ | 0.370 | 0.309 | 0.243 | 0.192 | 0.168 | 0.136 | 0.124 | 0.113 |

The rate equation is given by:
$$Rate = k[\mathbf{I}]^n[\mathbf{II}]^n$$
(a) Plot the concentration of compound **II** against time. What information can you deduce from this plot? (b) What is the overall order of the reaction? (c) Determine the rate constant.
[Data based on: M.G. Silvestri and C.E. Dills, *Journal of Chemical Education* (1989) vol. 66, p. 690.]

9.  The oxidation of $H_2$ by permanganate ion is catalysed by nitrate ions. If the pressure of $H_2$ is 1 bar and the permanganate ion is in vast excess, the rate equation can be written in the form:
    $$Rate = k_{obs}[NO_3^-]^n$$
    Using the initial rates method, the following kinetic data were obtained:

| [NO$_3^-$]$_0$ / mol dm$^{-3}$ | 0.014 | 0.029 | 0.044 | 0.058 | 0.072 | 0.087 |
|---|---|---|---|---|---|---|
| log (initial rate) | -2.35 | -2.10 | -1.88 | -1.78 | -1.73 | -1.54 |

(a) What is the value of $n$ in the rate equation? (b) What do the results tell you about the role of the catalyst in this reaction?
[Data based on: H. Nechamkin, E. Keller and J. Goodkin, *Journal of Chemical Education* (1977) vol. 54, p. 775.]

10.    For a reaction of the general type:
$$A + B \rightarrow products$$
the rate equation is of the form:
$$Rate = k[A]^n[B]^m$$
For a specific reaction of this type, the change in concentration of A was monitored with reactant B in vast excess and the results are given below:

| Time $\times 10^{-4}$ / s | 0 | 1 | 5 | 10 | 20 | 30 | 40 |
|---|---|---|---|---|---|---|---|
| [A] / mol dm$^{-3}$ | 0.1 | 0.091 | 0.065 | 0.048 | 0.031 | 0.023 | 0.018 |

The experiment was repeated several times with different initial concentrations of A, and the initial rate of reaction was found for each run. These data were as follows:

| $[A]_0$ / mol dm$^{-3}$ | 0.1 | 0.08 | 0.04 | 0.02 | 0.01 |
|---|---|---|---|---|---|
| Initial rate $\times 10^8$ / mol dm$^{-3}$ s$^{-1}$ | 105.0 | 68.0 | 18.0 | 5.0 | 1.5 |

(a) Plot a graph of [A] against time for an initial concentration $[A]_0 = 0.1$ M and determine the initial rate. Compare your answer with that given in the second table of data. (b) Use the initial rate data to determine the order of the reaction with respect to A assuming that $n$ is an integral value. (c) Determine a value of the rate constant for the reaction. Does this value correspond to $k$ in the rate equation given above?
[Data based on: J. Casado, M.A. López-Quintela and F.M. Lorenzo-Barrel, *Journal of Chemical Education* (1986) vol. 63, p. 450.]

11.    The cobalt(III) complex cation $[Co(NH_3)_5(NO_2)]^+$ possesses two *linkage isomers* in which the nitrite ion is bonded to the cobalt(III) centre through either the oxygen or the nitrogen atom. Over a period of days, the isomer in which the $[NO_2]^-$ group is oxygen-bonded converts to that in which it is bound through nitrogen:
$$Isomer\ \mathbf{I} \rightarrow Isomer\ \mathbf{II}$$

➤

**Linkage isomers: see Section 16.5 in H&C**

The isomerization can be followed using infrared spectroscopy by monitoring the disappearance of an absorption band at 1600 cm$^{-1}$. Data are tabulated below, and the infinity reading ($t = \infty$) corresponds to the absorbance at 1600 cm$^{-1}$ when the isomerization is complete.

➤

**IR spectra may be recorded in terms of transmittance or absorbance: see Section 9.1 of H&C**

| Time / days | 0 | 0.125 | 0.5 | 2.0 | 4.0 | 11.0 | 18.0 | $\infty$ |
|---|---|---|---|---|---|---|---|---|
| Absorbance | 0.480 | 0.462 | 0.414 | 0.349 | 0.283 | 0.156 | 0.118 | 0.108 |

(a) Correct each absorbance value for the fact that the absorption at 1600 cm$^{-1}$ does not completely disappear at $t = \infty$. (b) Plot the corrected absorbance values, $A_{corr}$, against time. What information does the graph give you? (c) Determine the order of the reaction with respect to isomer **I**. (d) Determine the rate constant.
[Data from: W.H. Hohman, *Journal of Chemical Education* (1974) vol. 51, p. 553.]

12. Cerium(IV) ions oxidize iron(II) ions according to the equation:

$$Ce^{4+}(aq) + Fe^{2+}(aq) \rightarrow Ce^{3+}(aq) + Fe^{3+}(aq)$$

When this reaction is carried out as a redox titration, a suitable indicator is N-phenylanthranilic acid, **10.1**; at the end-point the solution changes from colourless to purple ($\lambda_{max}$ = 580 nm). On standing after the titration, the purple colour fades to yellow and the kinetics of this decay can be followed by monitoring the absorbance at 580 nm. Typical data are given below:

| Time / s | 30 | 60 | 90 | 120 | 180 | 240 | 300 | 360 | 420 | 480 |
|---|---|---|---|---|---|---|---|---|---|---|
| Absorbance | 0.65 | 0.60 | 0.55 | 0.52 | 0.46 | 0.40 | 0.32 | 0.28 | 0.24 | 0.21 |

**(10.1)**

(a) Plot the absorbance, $A$, against time. How easy is it to state that the decay is *not* zero order? (b) Show that the decay follows first order kinetics. (c) Determine the rate constant.
[Data from: S.K. Mishra, P.D. Sharma and Y.K. Gupta, *Journal of Chemical Education* (1976) vol. 53, p. 327.]

## THE ARRHENIUS EQUATION

The Arrhenius equation (equation 10.5) relates the rate constant to the temperature of a reaction, although *the equation is not always obeyed over wide ranges of temperature.*

$$\ln k = \ln A - \frac{E_{activation}}{RT} \tag{10.5}$$

where:    $k$ = rate constant
$R$ = molar gas constant = $8.314 \times 10^{-3}$ kJ K$^{-1}$ mol$^{-1}$
$T$ = temperature (K)
$E_{activation}$ = activation energy (kJ mol$^{-1}$)
$A$ = frequency factor (discussed in Section 10.6 of H&C)

**Problem set 10.2**

1. If the rate constants for a reaction are measured at two temperatures, a value of $E_{activation}$ could be estimated from the equation:

➤
**Derived in H&C,
Section 10.6**

$$\ln \frac{k_2}{k_1} = \frac{E_{activation}}{R} \left[ \frac{1}{T_1} - \frac{1}{T_2} \right]$$

Why is it preferable to measure values of $k$ at more than two temperatures in order to determine $E_{activation}$?

2. Following from your answer to question 1, the data in the following table show how values of the first-order rate constant $k$ for the decomposition of an organic peroxide vary with temperature.

| Temperature / K | 410 | 417 | 426 | 436 |
|---|---|---|---|---|
| $k$ / s$^{-1}$ | 0.0193 | 0.0398 | 0.0830 | 0.2170 |

Use pairs of readings for $T = 410$ and 417, 417 and 426, and 417 and 436 to determine three separate values of $E_{activation}$. Compare your values with that obtained in worked example 10.6 in H&C by a graphical method. What conclusions can you draw?

➤ 3. The solvolysis of 2-chloro-2-methylpropane is first order with respect to the halogenoalkane. The first-order rate constant varies with temperature as shown in the table below. Determine a value of $E_{activation}$ for the reaction.

**Solvolysis is the reaction of a substrate with solvent; hydrolysis is the specific case when the solvent is water**

| Temperature / K | 288 | 298 | 308 |
|---|---|---|---|
| $k \times 10^5$ / s$^{-1}$ | 2.78 | 8.59 | 26.1 |

[Data from: J.A. Duncan and D.J. Pasto, *Journal of Chemical Education* (1975) vol. 52, p. 666.]

4. The reaction of pyridine with the bromo-compound **I** gives a pyridinium salt and can be followed by measuring changes in the conductance of the solution.

**I**        pyridine

(a) Why can the rate of reaction be monitored by taking conductance measurements? (b) Use the data below to determine the activation energy for the reaction. (c) Estimate a value of the rate constant at 298 K.

| Temperature / K | 293.2 | 296.1 | 299.7 | 301.0 | 305.9 | 307.8 | 310 | 315.8 |
|---|---|---|---|---|---|---|---|---|
| $k \times 10^2$ / dm$^3$ mol$^{-1}$ min$^{-1}$ | 1.36 | 2.03 | 2.75 | 2.79 | 4.07 | 4.75 | 5.27 | 7.73 |

[Data from: P.W.C. Barnard and B.V. Smith, *Journal of Chemical Education* (1981) vol. 58, p. 282.]

## ELEMENTARY STEPS AND THE STEADY-STATE APPROXIMATION

The following problems test basic concepts and do **not** assume that you have studied the derivation in Box 10.4 of H&C.

**Problem set 10.3**

1.  Which of the following steps are unimolecular and which are bimolecular?

    (a) $H_2 \rightarrow 2H^\bullet$

    (b) $H^\bullet + Br_2 \rightarrow HBr + Br^\bullet$

    (c) $2Cl^\bullet \rightarrow Cl_2$

    (d) $Me_3CBr \rightarrow [Me_3C]^+ + Br^-$

2.  Write a rate equation for each of the elementary steps shown in question 1.

3.  Some reaction steps that are proposed as being involved in the depletion of the ozone layer are:

    $$2ClO^\bullet \rightarrow Cl_2O_2$$
    $$Cl_2O_2 \xrightarrow{h\nu} 2Cl^\bullet + O_2$$
    $$Cl^\bullet + O_3 \rightarrow ClO^\bullet + O_2$$

    (a) Write a rate equation for each step, assuming that the rate constant for each can be represented as $k$. (b) Which steps are bimolecular?

4.  What relationships, if any, do the reaction orders with respect to specific reactants have to the stoichiometry of the equation for (a) an elementary step, and (b) an overall reaction?

5.  The mechanism of the decomposition reaction:

    $$A \rightarrow B + C$$

    can be written as a sequence of elementary steps:

    Step **I**:      $A \xrightarrow{k_1} D$

    Step **II**:     $D \xrightarrow{k_2} B + C$

    where D is an intermediate species.

    (a) For step **I** alone, give an expression that shows the rate of appearance of D. (b) How is your answer to part (a) influenced if you take into account the fact that D is being consumed in step **II**? (c) Use the steady-state approximation to show that the rate of formation of B is directly proportional to the concentration of A.

# 11 Hydrogen and the *s*-block elements

---

**Topics**

- Acid dissociation constants
- Base dissociation constants and the relationship between p$K_a$ and p$K_b$
- pH calculations
- Acid-base titrations
- Indicators
- Buffers

---

In Chapter 11 of the main text, we described some of the chemistry of dihydrogen and the elements of groups 1 and 2. Section 11.9 dealt with Brønsted acids and bases, and this topic is the focus of attention in this chapter of the workbook. You should refer to worked examples 11.1 to 11.8 in H&C for guidance when answering the problem sets 11.1 to 11.3 below.

---

## ACID DISSOCIATION CONSTANTS

**Problem set 11.1**

1. Write equations to show the dissociation in aqueous solution of the following acids: (a) $CH_3CO_2H$; (b) $HNO_3$; (c) $H_2SO_3$. [*Hint*: refer to Table 11.3 on page 535 of H&C.]

2. The value of $K_a$ for nitrous acid, $HNO_2$, is $4.7 \times 10^{-4}$ mol dm$^{-3}$. Find p$K_a$.

3. The value of p$K_a$ for formic acid, $HCO_2H$, is 3.75. Determine $K_a$.

4. Citric acid, structure **11.1**, is characterized by three p$K_a$ values: 3.14, 4.77 and 6.39. Which protons dissociate in aqueous solution?

5. The p$K_a$ values of acetic acid and chloroacetic acid are 4.77 and 2.85, respectively. Which is the stronger acid in aqueous solution?

6. The values of $K_a$ for hypobromous acid (HOBr) and hypochlorous acid (HOCl) are $2.1 \times 10^{-9}$ and $3.0 \times 10^{-5}$ mol dm$^{-3}$ respectively. Which is the stronger acid in aqueous solution?

7. Determine the number of moles of hydrogen ions in 25 cm$^3$ of a 0.20 M solution of hydrochloric acid.

8. Determine the concentration of hydrogen ions in a 0.20 M solution of acetic acid ($K_a = 1.7 \times 10^{-5}$ mol dm$^{-3}$).

9. Determine the concentration of cyanide ions in a 0.050 M solution of hydrocyanic acid, HCN (p$K_a = 9.40$).

10. What is the relationship between the concentrations of chloride and hydrogen ions in a 0.1 M aqueous solution of hydrochloric acid?

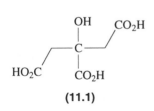

**(11.1)**

## BASE DISSOCIATION CONSTANTS AND THE RELATIONSHIP BETWEEN p$K_a$ AND p$K_b$

Whilst the dissocation properties of weak acids are quoted in terms of $K_a$ and p$K_a$, you may find the degree of dissociation of a weak base described in terms of $K_b$, p$K_b$, p$K_a$ or $K_a$. The relationship between p$K_b$ and p$K_a$ is given in equation 11.1 and allows you to determine both the concentration of hydroxide and hydrogen ions in a basic solution (equation 11.2).

> $K_w$ = self-ionization constant of water

$$pK_a + pK_b = pK_w = 14.00 \tag{11.1}$$

$$K_w = [H^+][OH^-] = 1 \times 10^{-14} \ mol^2 \ dm^{-6} \tag{11.2}$$

**Problem set 11.2**

1. Caesium hydroxide is a strong base. (a) How many moles of $[OH]^-$ ions are present in 30 cm$^3$ of a 0.40 M aqueous solution of CsOH? (b) What volume of 0.20 M nitric acid is required to completely neutralize the CsOH present?
2. Determine $K_b$ for a base for which p$K_b$ = 5.2.

> **Pyridine:** see Section 15.12 in H&C

3. Determine p$K_b$ for pyridine if $K_b = 1.78 \times 10^{-9} \ mol \ dm^{-3}$
4. (a) Write an equation to show the equilibrium that is set up when ammonia dissolves in water. (b) Determine the concentration of $[OH]^-$ ions present in a 0.150 M aqueous ammonia solution (p$K_b$ = 4.75).
5. Three bases are characterized by the following dissociation constants: base **A**, p$K_b$ = 9.10, base **B**, $K_b = 7.10 \times 10^{-11} \ mol \ dm^{-3}$, base **C**, p$K_a$ = 5.12. Place these compounds in order of *increasing* base strength in aqueous solution.

## pH CALCULATIONS

Equation 11.3 is the all important equation for this section. However, you must bear in mind whether the acid or base with which you are dealing is fully or partially dissociated in aqueous solution.

$$pH = -\log \ [H^+] \tag{11.3}$$

**Worked example 11.1**

If 25 cm$^3$ of 0.20 M aqueous nitric acid is added to 20 cm$^3$ 0.18 M sodium hydroxide, what is the pH of the resulting solution?

Firstly, write a balanced equation for the *neutralization* reaction:

$$HNO_3(aq) + NaOH(aq) \rightarrow NaNO_3(aq) + H_2O(l)$$

Determine the number of moles of $HNO_3(aq)$ used, the number of moles of NaOH(aq) used and the number of moles of excess reagent.

$$\text{Moles } H^+ = \frac{25 \times 0.20}{1000} = 5.0 \times 10^{-3}$$

> **Strictly, we should write $[H_3O]^+$ to signify hydrogen ion in aqueous solution, but $H^+$ is often used**

$$\text{Moles } [OH]^- = \frac{20 \times 0.18}{1000} = 3.6 \times 10^{-3}$$

$HNO_3$ and NaOH react in a 1:1 molar ratio; the acid is in excess:

$$\text{Moles of excess } H^+ = (5.0 \times 10^{-3}) - (3.6 \times 10^{-3}) = 1.4 \times 10^{-3}$$

Now note the *total* volume of solution (45 cm³), calculate the concentration of excess $H^+$ ions, and so determine the pH of the solution:

$$[H^+] = \frac{1.4 \times 10^{-3} \times 1000}{45} = 0.031$$

$$pH = -\log [H^+] = 1.5$$

---

**Problem set 11.3**

1.  (a) Determine the pH of a 0.10 M aqueous solution of hydrochloric acid. (b) Are the pH values of 25 cm³ 0.10 M and 50 cm³ 0.10 M solutions of hydrochloric acid the same?
2.  What is the pH of a 1 dm³ aqueous solution that contains 2 g of dissolved sodium hydroxide? [$A_r$ Na = 23; O = 16; H = 1].
3.  Find the pH of 0.050 M aqueous acetic acid ($K_a = 1.7 \times 10^{-5}$ mol dm⁻³).
4.  What is the *change* in pH upon diluting 0.1 M hydrochloric acid with water by a factor of ten?
5.  Determine the pH of an aqueous solution of 0.25 M sulfuric acid.
6.  If 35 cm³ of 0.10 M aqueous hydrochloric acid are added to 30 cm³ 0.10 M potassium hydroxide, what is the pH of the resulting solution?
7.  (a) Write a balanced equation for the reaction between barium hydroxide and nitric acid. (b) What is the pH of the resulting solution after 20 cm³ 0.010 M nitric acid has been added to 20 cm³ 0.010 M aqueous barium hydroxide?
8.  Determine the pH of a solution of 8 cm³ 0.25 M acetic acid to which 10 cm³ 0.25 M sodium hydroxide has been added.
9.  Determine the pH of a 0.10 M aqueous solution of ammonia ($pK_b = 4.75$).
10. (a) Write an equation for the reaction between aqueous ammonia and nitric acid. (b) Calculate the pH of a 25 cm³ 0.080 M aqueous solution of ammonia to which 25 cm³ 0.10 M nitric acid has been added.

---

## ACID-BASE TITRATIONS

The questions above included examples of neutralization reactions in which the pH of the final solution depended upon the reagent that was in excess. During an acid-base *titration*, the acid (or the base) is added from a graduated *burette* to the base (or acid) contained in a conical flask. The reaction can be monitored by
   • measuring the pH of the solution in the flask using a pH meter, or
   • using an acid-base indicator that changes colour at the point of neutralization (the *end-point*).
   In this section we consider the changes of pH during a titration, and in Section 11.5 we look at the use of acid-base indicators.

### Interpretation of pH titration curves

Figure 11.1 shows the change in pH during the addition of 30 cm³ of 0.10 M NaOH(aq) to 25 cm³ 0.10 M HCl(aq). Initially, the flask contains a *fully dissociated acid*, and you should show that initially the pH is 1.0. The base is also fully dissociated in aqueous solution. Fill in the chart below and check that the values you obtain correspond to the values plotted in Figure 11.1.

| Volume of 0.10 M NaOH(aq) added / cm³ | 1.0 | 5.0 | 10.0 | 20.0 | 22.0 | 24.5 | 24.9 |
|---|---|---|---|---|---|---|---|
| Total volume | | | | | | | |
| pH of solution | | | | | | | |

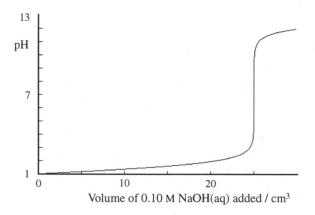

**11.1** The variation in pH during the addition of 30 cm³ of 0.10 M NaOH(aq) to 25 cm³ 0.10 M HCl(aq).

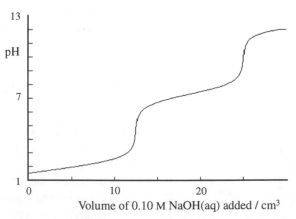

**11.2** The variation in pH during the addition of 30 cm³ of 0.10 M NaOH(aq) to 25 cm³ 0.10 M CH₃CO₂H(aq).

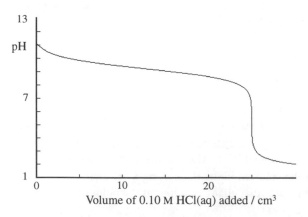

**11.3** The variation in pH during the addition of 30 cm³ of 0.10 M HCl(aq) to 25 cm³ 0.10 M NH₄OH(aq).

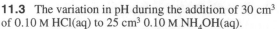

**11.4** The variation in pH during the addition of 30 cm³ of 0.10 M NaOH(aq) to 25 cm³ 0.10 M H₂SO₃(aq).

The *end-point* of the titration is the point at which the alkali added has exactly neutralized the acid — neither reagent is in excess. On either side of the end-point (pH 7.0), notice how sensitive the pH is to the composition of the solution.

Figure 11.2 shows the variation in pH during the addition of 0.10 M NaOH(aq) to 0.10 M acetic acid ($pK_a = 4.77$). Initially, the flask contains a weak acid; confirm that the initial pH is 2.88. Compare the shapes of the initial parts of the plots in figures 11.1 and 11.2; whereas the addition of NaOH(aq) to HCl(aq) caused little change in pH during the very first part of the titration, a more significant change is seen when NaOH(aq) is added to $CH_3CO_2H$(aq). A second point of significance is the fact that the pH at the end-point in Figure 11.2 is *not* at 7.0 as it was in the HCl-NaOH titration. In the case of a strong base added to a weak monobasic acid, the end-point is in the range $7 < pH_{\text{end-point}} < 14$. The value depends upon the $pK_a$ of the weak acid.

The neutralization reaction in this example is:

$$CH_3CO_2H(aq) + NaOH(aq) \rightarrow Na[CH_3CO_2](aq) + H_2O(l)$$
weak acid             salt of a weak acid

A solution that contains a weak acid and a salt of the same acid is called a *buffer solution*, and we look at such systems in Section 11.6.

Figure 11.3 shows the change in pH as 0.10 M hydrochloric acid is added to 0.10 M aqueous ammonium hydroxide. For $NH_3$, $pK_b = 4.75$ — confirm that the initial pH of the solution is 11.1. The effect of the neutralization reaction can be seen from the curve in Figure 11.3; notice that the end-point for the titration between a weak base and a fully dissociated acid lies in the range $7 > pH_{\text{end-point}} > 0$.

Figure 11.4 follows the titration between a fully dissociated base and a weak acid but the graph has a quite different profile from that in Figure 11.2. Acetic acid (Figure 11.2) is monobasic but sulfurous acid (Figure 11.4) is dibasic ($pK_a = 1.82$ and $6.92$). Two separate neutralization steps take place:

$$H_2SO_3(aq) + NaOH(aq) \rightarrow NaHSO_3(aq) + H_2O(l)$$

$$NaHSO_3(aq) + NaOH(aq) \rightarrow Na_2SO_3(aq) + H_2O(l)$$

## INDICATORS

**(11.2)**

Acid-base indicators such as phenolphthalein, **11.2**, and bromocresol green can be considered as weak acids of the general type HIn in which the conjugate base $[In]^-$ has a *different* colour from the acid HIn (equation 11.4).

$$HIn(aq) + H_2O(l) \rightleftharpoons [H_3O]^+(aq) + [In]^-(aq) \tag{11.4}$$
colour I                  colour II

By Le Chatelier's principle, if the indicator is in an *acidic* solution, equilibrium 11.4 shifts to the left-hand side and if the solution is alkaline, the equilibrium shifts to the right-hand side.

Consider the acid dissociation constant $K_a$ for an indicator HIn (equation 11.5).

$$K_a = \frac{[H_3O^+][In^-]}{[HIn]}$$

(11.5)

The colour (**I** or **II** in equation 11.4) of the solution depends upon the ratio of concentrations $[In^-] : [HIn]$, obtained by rearranging equation 11.5 to give equation 11.6. This is usually expressed in the logarithmic form shown in equation 11.7.

$$\log \frac{x}{y} = \log x - \log y$$

$$\frac{K_a}{[H_3O^+]} = \frac{[In^-]}{[HIn]}$$

(11.6)

$$\log K_a - \log [H_3O^+] = -pK_a + pH = \log \frac{[In^-]}{[HIn]}$$

(11.7)

---

| Worked example 11.2 |
|---|

**Phenolphthalein ($pK_a$ = 9.50) is colourless and its conjugate base is pink. In what pH range will an aqueous solution of phenolphthalein change colour?**

From equation 11.7 we can write:

$$-9.50 + pH = \log \frac{[In^-]}{[HIn]}$$

We could randomly test pH values but you should notice the following important result. At pH = 9.5:

$$-9.50 + 9.50 = 0 = \log \frac{[In^-]}{[HIn]} \qquad \frac{[In^-]}{[HIn]} = 1.00$$

The colour will therefore change around this region:
At pH 8.0:

$$-9.50 + 8.00 = -1.50 = \log \frac{[In^-]}{[HIn]} \qquad \frac{[In^-]}{[HIn]} = 0.03$$

At pH 10.0:

$$-9.50 + 10.00 = 0.50 = \log \frac{[In^-]}{[HIn]} \qquad \frac{[In^-]}{[HIn]} = 3.16$$

The colourless $\rightarrow$ pink colour change occurs when the dominant solution species changes from being HIn to $[In]^-$. At pH 8.00, $[HIn] \approx 33 \times [In^-]$ and the solution is colourless. At pH 10.00, the solution is observed to be pink. The pH range from 8 to 10 may seem to be rather imprecise but Figure 11.1 helps you to understand why phenolphthalein is a suitable indicator for a strong acid-strong base titration. Mark the pH range 8–10 on the curve; you should see that the quantity of alkali needed to cause this pH change (and thus an indicator colour change) is very small. That is, the end-point accurately corresponds to the volume of NaOH(aq) added at the instant the colourless $\rightarrow$ pink change occurs.

---

Table 11.1 lists some acid-base indicators. You should note that for titrations involving weak acids with weak bases (e.g. $CH_3CO_2H$ against aqueous $NH_3$), the end-point does *not* correspond to a large change in pH. Consequently, it is not possible to follow such titrations using an indicator because the indicator changes colour over the period that a relatively large volume of acid or base is added. For an accurate titration result, a sharply defined end-point is essential.

**Problem set 11.4**

1.  Using Table 11.1, determine which indicators would be suitable for the titrations shown in (a) Figure 11.1, (b) Figure 11.2, and (c) Figure 11.3.
2.  What problem is encountered when choosing an indicator for the titration between $H_2SO_3$ and NaOH (Figure 11.4)?
3.  Calculate the ratio [HIn] : [In⁻] at pH 5.00 for HIn being (a) methyl orange, and (b) phenol red.
4.  What do the data in Table 11.1 tell you about the basicity of thymol blue?

**Table 11.1**    Selected acid-base indicators in aqueous solution

| Indicator | $pK_a$ | Colour change (acid → basic solution) | pH range in which indicator colour changes |
|---|---|---|---|
| Phenolphthalein | 9.50 | Colourless → pink | 8.0–10.0 |
| Phenol red | 8.00 | Yellow → red | 6.8–8.2 |
| Bromocresol purple | 6.40 | Yellow → purple | 5.2–6.8 |
| Bromocresol green | 4.90 | Yellow → blue | 3.8–5.4 |
| Methyl orange | 3.46 | Red → yellow | 3.2–4.4 |
| Thymol blue | 1.65 | Red → yellow | 1.2–2.8 |
| | 9.20 | Yellow → blue | 8.0–9.8 |

## BUFFERS

### What is a buffer and how does it work?

A *buffer* is a solution to which reasonable amounts of acid or base can be added without causing a significant change in the pH. A buffer usually consists of a solution of a weak acid and a salt of that acid, (e.g. acetic acid and sodium acetate) or a weak base and its salt. Buffers are very important in living organisms where a constant pH is essential; human blood has a pH of 7.4 and is naturally *buffered*.

Let us consider how a buffer works. Equations 11.8 and 11.9 show that in aqueous solution, $CH_3CO_2H$ is partially dissociated whilst its sodium salt is fully dissociated.

$$CH_3CO_2H(aq) + H_2O(l) \rightleftharpoons [H_3O]^+(aq) + [CH_3CO_2]^-(aq) \tag{11.8}$$

$$Na[CH_3CO_2](s) \xrightarrow{\text{water}} Na^+(aq) + [CH_3CO_2]^-(aq) \tag{11.9}$$

If a small amount of acid is added to the buffer, it will be consumed by $[CH_3CO_2]^-$ to form more $CH_3CO_2H$ in the reverse of reaction 11.8. Any $[OH]^-$ ions added to the buffer will be neutralized by $[H_3O]^+$, and by Le Chatelier's principle, equilibrium 11.8 will shift to the right-hand side to re-establish the original concentration of hydrogen ions in solution.

### Determining the pH of a buffer

Consider 500 cm$^3$ of a solution of 0.1 M $CH_3CO_2H$ ($pK_a = 4.77$). From equation 11.8, we know that the concentration of hydrogen ions and acetate ions are equal and small. Now add 500 cm$^3$ 0.1 M $Na[CH_3CO_2]$ to the acetic acid; the total volume is 1 dm$^3$. From equation 11.9, we know that sodium acetate is fully dissociated. Thus, it is reasonably valid to make the approximation that the *total* concentration of $[CH_3CO_2]^-$ ions is equal to the *concentration of the salt*, since far fewer acetate ions originate from the acid than from the salt.

Equation 11.10 gives the expression for $K_a$ for $CH_3CO_2H$.

$$K_a = \frac{[H_3O^+][CH_3CO_2^-]}{[CH_3CO_2H]} \tag{11.10}$$

For the buffer solution, we have already said that $[CH_3CO_2^-]$ is *approximately* equal to the initial concentration of the salt $Na[CH_3CO_2]$. For weak acids, we usually make the assumption that $[CH_3CO_2H]_{equilibrium} \approx [CH_3CO_2H]_{initial}$. Thus, we can rewrite equation 11.10 in terms of known acid and salt concentrations (equation 11.11) and by taking logarithms of both sides of the equation, we obtain expression 11.12 which links the values of $pK_a$, pH and the initial concentrations of acid and salt.

$$K_a = \frac{[H_3O^+][salt]}{[acid]} \tag{11.11}$$

➤
**Compare with equations
11.6 and 11.7**

$$\log K_a = \log [H_3O^+] + \log \frac{[salt]}{[acid]}$$

$$pH - pK_a = \log \frac{[salt]}{[acid]} \tag{11.12}$$

---

| Worked example 11.3 |
|---|

Determine the pH of a buffer which is 0.20 M with respect to both $CH_3CO_2H$ ($pK_a = 4.77$) and $Na[CH_3CO_2]$.

$$pH - pK_a = \log \frac{[salt]}{[acid]}$$

$$pH - 4.77 = \log \frac{[0.2]}{[0.2]}$$

$$pH - 4.77 = \log 1 = 0$$

$$pH = 4.77$$

---

# 12 Thermodynamics and electrochemistry

> **Topics**
> • Internal energy and work done
> • Heat capacities and Kirchhoff's equation
> • Free energy and the reaction isotherm
> • Free energy and entropy
> • Electrochemistry
> • Solubility product constant

## INTERNAL ENERGY AND WORK DONE

The First Law of Thermodynamics states that energy cannot be created or destroyed, merely changed from one form to another. In this section we are concerned with the change in internal energy, $\Delta U$, of a system during a reaction; the total energy of the system *and* the surroundings remains constant but the internal energy of the system itself may change (equation 12.1).

**Refer to worked examples 12.1 and 12.2 in H&C**

$$\Delta U = q + w = \Delta H + w \qquad\qquad w = \text{work done} \qquad (12.1)$$

The enthalpy change $\Delta H$ may be positive (heat energy is transferred to the system from the surroundings) or negative (heat energy is transferred from the system to the surroundings).

The most common example that we shall encounter of *work done* is the expansion or compression of a gas; if work is done *by* the system *on* the surroundings (e.g. a gas expands), it is defined as negative work (equation 12.2).

$$\text{Work done } by \text{ the system } on \text{ the surroundings} = w = -P\Delta V \qquad (12.2)$$

**1 bar = $10^5$ Pa**

where $w$ is in J $\quad$ $P$ = pressure (Pa) $\quad$ $\Delta V$ = change in volume ($m^3$)

Combining equations 12.1 and 12.2 gives equation 12.3.

$$\Delta U = \Delta H - P\Delta V \qquad\qquad \Delta U \text{ and } \Delta H \text{ in joules} \qquad (12.3)$$

In *all* the problems below, assume that apparatus used to collect gases is frictionless. [Why is this assumption necessary?]

**Problem set 12.1**

**Molar gas constant, $R = 8.314$ J mol$^{-1}$ K$^{-1}$**
**Volume of 1 mole of gas (1 bar, 273 K) = 22.7 dm$^3$**

1. When dilute sulfuric acid is added to solid copper(II) carbonate, (a) is work done by or on the system, and (b) is the work done negative or positive?

2. Calculate the work done on the surroundings when 0.1 moles of $Na_2CO_3$ are decomposed by the addition of hydrochloric acid at 1 bar and 273 K.

3. When an air bag in a motor car inflates, is work done by or on the system?

4.    When 1 mole of hexane (mp 178 K, bp 342 K) reacts completely with dioxygen, is work (measured at 298 K) done by or on the surroundings?

➤ 5.    Calculate the work done by the system at 1 bar and 298 K when 2 moles of sodium azide $NaN_3$ decomposes to give sodium and dinitrogen.

**This reaction takes place in motor vehicle air bags**

6.    (a) Write an equation for the addition of $H_2$ to ethene. (b) Calculate the work done on the system at 2 bar and 300 K when 1.2 moles of ethene react with 1.2 moles of dihydrogen.

## HEAT CAPACITIES AND KIRCHHOFF'S EQUATION

### Molar heat capacities

Equations 12.4 and 12.5 define the molar heat capacity, $C$, at constant pressure or constant volume.

$$C_P = \left(\frac{\partial H}{\partial T}\right)_P \tag{12.4}$$

$$C_V = \left(\frac{\partial U}{\partial T}\right)_V \tag{12.5}$$

We showed in Section 12.3 of H&C that $C_P$ and $C_V$ are related by equation 12.6.

**$R$ = Molar gas constant**

$$C_P = C_V + R \qquad \text{for one mole of an ideal gas} \tag{12.6}$$

Here we look at the derivation of this relationship in more detail in order to understand why we were able to make the assumption that:

$$\left(\frac{\partial U}{\partial T}\right)_P = \left(\frac{\partial U}{\partial T}\right)_V \tag{12.7}$$

➤ **In order to follow this derivation, you should be familiar with partial differentials: see Box 12.1 in H&C**

The total differential $dU$ is given by expression 12.8; we differentiate $U$ with respect to the variables $T$ and $V$, one at a time, with the second variable held constant.

$$dU = \left(\frac{\partial U}{\partial T}\right)_V dT + \left(\frac{\partial U}{\partial V}\right)_T dV \tag{12.8}$$

Differentiating equation 12.8 with respect to $T$ at constant pressure gives equation 12.9. From this, we can deduce that for an ideal gas, equation 12.7 is true.

$$\left(\frac{\partial U}{\partial T}\right)_P = \left(\frac{\partial U}{\partial T}\right)_V\left(\frac{\partial T}{\partial T}\right)_P + \left(\frac{\partial U}{\partial V}\right)_T\left(\frac{\partial V}{\partial T}\right)_P \tag{12.9}$$

This term equals 1.    For an *ideal gas*, this term equals 0.

### Kirchhoff's equation

Kirchhoff's equation allows us to see to what extent $\Delta H$ remains constant over a given temperature range; the integrated form of Kirchhoff's relationship is given

➤

**Refer to Box 12.2 in H&C
for the derivation of
equation 12.10**

in equation 12.10. We often make the assumption that enthalpy changes have constant values but this is not necessarily valid.

$$\Delta H_{(T_2)} - \Delta H_{(T_1)} = \Delta C_P \times (T_2 - T_1) \qquad (12.10)$$

**Problem set 12.2**

1. Consider the reaction: $H_2(g) \rightarrow 2H(g)$.
   At 298 K, $\Delta_f H^\circ(H, g) = 218.0$ kJ mol$^{-1}$. Determine the value of this enthalpy change at 400 K if values of $C_P$ over this temperature range for $H_2(g)$ and $H(g)$ are 29.0 and 20.8 J K$^{-1}$ mol$^{-1}$ respectively.

2. Consider the reaction: $HF(g) \rightarrow H(g) + F(g)$
   At 298 K, $\Delta_f H^\circ(H, g) = 218.0$, $\Delta_f H^\circ(F, g) = 79.4$ and $\Delta_f H^\circ(HF, g) = -273.0$ kJ mol$^{-1}$. (a) Determine the value at $\Delta_r H^\circ(298$ K$)$ for the above reaction. (b) If values of $C_P$ for $H(g)$, $F(g)$ and $HF(g)$ are 20.8, 22.5 and 29.1 J K$^{-1}$ mol$^{-1}$ respectively, estimate the change in enthalpy for the reaction at 500 K.

3. (a) Write an equation for the combustion of gaseous CO. (b) Estimate the enthalpy change for this reaction at 320 K given the following data: $\Delta_f H^\circ(CO_2$, g, 298 K$) = -393.5$ kJ mol$^{-1}$, $\Delta_f H^\circ(CO$, g, 298 K$) = -110.5$ kJ mol$^{-1}$; $C_P$ (between 298–320 K) for $CO(g)$, $O_2(g)$ and $CO_2(g) = 29.2$, 29.4 and 37.2 J K$^{-1}$ mol$^{-1}$ respectively.

4. How valid is it to assume that $C_P$ remains constant for the following species over the temperature range 298 to 500 K? (a) $H_2(g)$; (b) $CClF_3(g)$; (c) $HI(g)$; (d) $NO_2(g)$; (e) $NO(g)$; (f) $C_2H_4(g)$.

## FREE ENERGY AND THE REACTION ISOTHERM

### Changes in free energy

**Problem set 12.3**

1. Determine $\Delta_r G^\circ(298$ K$)$ for the following reactions:

   (a) $C_2H_2(g) + 2H_2(g) \rightarrow C_2H_6(g)$

   (b) $C_2H_4(g) + HCl(g) \rightarrow CH_3CH_2Cl(g)$

   (c) $2CO(g) + O_2(g) \rightarrow 2CO_2(g)$

   (d) $C_2H_5OH(l) \rightarrow H_2O(l) + C_2H_4(g)$

   (e) $CaCO_3(s) \rightarrow CaO(s) + CO_2(g)$

   (f) $Mg(s) + 2HCl(g) \rightarrow MgCl_2(s) + H_2(g)$

➤

**Values of $\Delta_f G^\circ(298$ K$)$ for
question 1 are listed in
Appendix 11 of H&C**

2. Which reactions in question 1 are thermodynamically favourable at 298 K?

3. What is the value of (a) $\Delta_f G^\circ(O_2$, g, 500 K$)$, and (b) $\Delta_f G^\circ(O_2$, g, 300 K$)$, (c) $\Delta_f H^\circ(O_2$, g, 300 K$)$,?

4. The value of $\Delta_f G^\circ(298$ K$)$ for $OF_2$ is +42 kJ mol$^{-1}$. Will $OF_2$ form when $F_2$ and $O_2$ combine at 298 K?

5. The Ellingham diagram shown in Figure 12.8 in H&C can be used to predict which metal oxides will be reduced by carbon. (a) At 1100 K, which metal oxides will carbon reduce? (b) Is it possible to reduce NiO with carbon at 500 K? (c) Can carbon be used to extract aluminium from alumina ($Al_2O_3$) below 1750 K?

**12.1** The dependence of $\Delta_f G^\circ$ on temperature for gaseous nitrogen dioxide, carbon monoxide, sulfur dioxide and germanium tetrachloride.

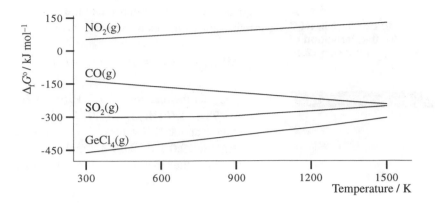

6.    Figure 12.1 shows how values of $\Delta_f G^\circ$ vary with temperature for four gaseous compounds. (a) Which compounds become less thermodynamically favoured as the temperature is raised from 300 to 1500 K? (b) For which compounds is the formation from their constituent elements favourable between 300 and 1500 K? (c) For which compound is $\Delta_f G^\circ$ approximately temperature invariant between 300 and 800 K? (d) Which compound is the most thermodynamically favoured at 900 K? (e) What can you say about the *relative* thermodynamic stabilities of $SO_2(g)$ and $CO(g)$ with respect to their constituent elements as a function of temperature? (f) Does the graph tell you anything about the relative stability of $NO_2$ with respect to dimerization to $N_2O_4$?

➤ **The reaction isotherm**

See pages 589-591 in H&C for a discussion concerning when the reaction isotherm applies to $K_p$ and $K_c$, or specifically to $K_p$

Equation 12.11 states the *reaction isotherm*. This important equation relates the equilibrium constant, $K$, for a reaction to the change in free energy.

$$\Delta G^\circ = -RT \ln K \tag{12.11}$$

**Problem set 12.4**

1.    (a) Determine $K_p$ for the formation of NO from $O_2$ and $N_2$ at 298 K if $\Delta_f G^\circ$ = +88 kJ per mole of NO. (b) What does the result tell you about the position of equilibrium: $^1/_2 N_2(g) + ^1/_2 O_2(g) \rightleftharpoons NO(g)$?

2.    The value of $K_p$ for the equilibrium:    $^1/_2 H_2(g) + ^1/_2 I_2(g) \rightleftharpoons HI(g)$
      is 0.5 at 298 K and 6.4 at 800 K. At each temperature, comment on the position of the equilibrium and determine $\Delta_f G^\circ$.

$R = 8.314 \times 10^{-3}$ kJ K$^{-1}$ mol$^{-1}$

3.    The following data refer to the formation of hydrocarbons from elements in their standard states at a temperature $T$ K. Use the data to determine $T$.

| Compound (gaseous) | $C_2H_2$ | $C_2H_4$ | $C_2H_6$ | $CH_4$ |
|---|---|---|---|---|
| $\Delta_f G^\circ(T$ K$)$ / kJ mol$^{-1}$ | +211 | +68 | −32 | −50 |
| $K$ (dimensionless) | $1.7 \times 10^{-37}$ | $1.0 \times 10^{-12}$ | $4.0 \times 10^5$ | $7.1 \times 10^8$ |

4.   (a) Determine $\Delta_f G^\circ$(298 K) for the reaction: $H_2O_2(l) \rightleftharpoons H_2O(l) + \frac{1}{2}O_2(g)$ using standard free energies of formation tabulated in Appendix 11 in H&C. (b) Hence determine $K$ for this system. (c) Why does $H_2O_2$ not decompose spontaneously on standing?

➤ 5.   The value of log $K_p$ for the formation of $SO_2(g)$ at 400 K is 39.3. Determine $\Delta_f G^\circ$(400 K).

*Care!* Tables of data may give either log *K* or ln *K*

6.   (a) The value of $\Delta_f G^\circ$(298 K) for the formation of HBr(g) is –53.4 kJ per mole of HBr. Determine $K_p$. (b) Is the formation of HBr from $H_2$ and $Br_2$ favoured at 298 K?

## FREE ENERGY AND ENTROPY

### Absolute entropy values and changes in entropy

The standard molar entropy of a substance varies with temperature according to equation 12.12 where it is assumed that $C_P$ is constant over the temperature range in question. [How valid is this assumption? — see Figure 12.5 in H&C.]

$$\Delta S = S_{(T_2)} - S_{(T_1)} = C_P \times \ln\left(\frac{T_2}{T_1}\right) \tag{12.12}$$

**Problem set 12.5**

**Refer to Appendix 11 in H&C for additional data needed for this problem set**

1.   The molar entropy $S^\circ$ of HCl at 298 K is 187 J K$^{-1}$ mol$^{-1}$. Determine $S^\circ$(350 K) if $C_P$ over this temperature range is 29.1 J K$^{-1}$ mol$^{-1}$.

2.   Which value is larger: $S^\circ(H_2O, l, 373 K)$ or $S^\circ(H_2O, g, 373 K)$?

3.   Values of $S^\circ$(Cu, s) at 400 K and 500 K are 40.5 and 46.2 J K$^{-1}$ mol$^{-1}$. Estimate the molar heat capacity of copper over this temperature range.

4.   The boiling point of $CCl_4$ is 349.5 K and $\Delta_{vap}H$ is 29.8 kJ mol$^{-1}$. What is the change in entropy when one mole of $CCl_4$ vaporizes at 1 bar?

5.   (a) What is the change in entropy when one mole of $H_2O$ vaporizes at 1 bar pressure (bp = 373 K, $\Delta_{vap}H = 40.6$ kJ mol$^{-1}$)? (b) Comment on the difference in the values obtained for $\Delta_{vap}S$ for $CCl_4$ (question 4) and $H_2O$.

➤ 6.   Figure 12.2 shows the variation in $S^\circ$ for germanium between 300 and 1500 K. Rationalize the shape of the curve.

**Structure of germanium at 298 K: see Section 7.10 in H&C**

7.   Determine the change in entropy at 298 K during the following reactions:
(a) $2Fe(s) + 3Cl_2(g) \rightarrow 2FeCl_3(s)$
(b) $Fe(s) + 2HCl(g) \rightarrow FeCl_2(s) + H_2(g)$
(c) $CaCO_3(s) \rightarrow CaO(s) + CO_2(g)$          ........continued

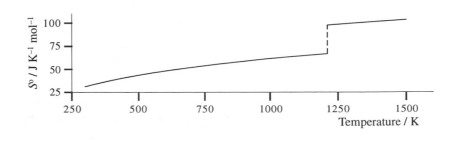

**12.2**  The variation in standard molar entropy, $S^\circ$, of germanium as a function of temperature over the range 300 to 1500 K.

(d) $C_2H_5OH(l) \rightarrow C_2H_4(g) + H_2O(l)$

(e) $H_2(g) + Cl_2(g) \rightarrow 2HCl(g)$

8.    Which reactions in question 7 are *entropically favoured*? Relate your numerical answers to the nature of the reactants and products in the reactions.

## The relationship between $\Delta G$, $\Delta H$ and $\Delta S$ for a reaction

Although changes in enthalpy are often used to give an indication of the feasibility of a reaction, it is the change in free energy that really controls whether or not a reaction may occur. Equation 12.13 relates the two quantities and in this section we apply this equation to selected reactions.

$$\Delta G = \Delta H - T\Delta S$$

$$\Delta G^\circ = \Delta H^\circ - T\Delta S^\circ \qquad \text{under standard state conditions} \qquad \left.\right\} \quad (12.13)$$

Some points to think about as you work through the problems are listed below.

• Are the signs of $\Delta G$ and $\Delta H$ for a given reaction necessarily the same?

• How important is the change in entropy in controlling whether or not a reaction may be thermodynamically favoured?

• Is it the $T\Delta S$ term or $\Delta S$ that is important?

**Problem set 12.6**

**Data not given in the questions are listed in Appendix 11 of H&C**

1.    (a) Write an equation for the formation of $MnCl_2$ from its constituent elements in their standard states. (b) If $\Delta_f H^\circ(MnCl_2, s)$ and $\Delta_f G^\circ(MnCl_2, s)$ are $-481$ and $-440.5$ kJ mol$^{-1}$ (at 298 K) respectively, determine the value of $\Delta_f S^\circ(MnCl_2, s)$. (c) Compare the value of $\Delta_f S^\circ(MnCl_2, s)$ you have calculated with that obtained using values of $S^\circ$.

2.    Consider the reaction: $\qquad H_2(g) + F_2(g) \rightarrow 2HF(g)$
(a) Determine $\Delta S^\circ(298 \text{ K})$ for this reaction given that values of $S^\circ(298 \text{ K})$ for $H_2(g)$, $F_2(g)$ and $HF(g)$ are 131, 203 and 174 J K$^{-1}$ mol$^{-1}$ respectively.
(b) Is the reaction entropically favoured? (c) Do you expect the values of $\Delta_f H^\circ(HF, g)$ and $\Delta_f G^\circ(HF, g)$ to be very different from each other?

3.    At 298 K, the value of $\Delta_f H^\circ(CH_3NH_2, g)$ is $-22.5$ kJ mol$^{-1}$. If values of $S^\circ$ for C(graphite) and $CH_3NH_2(g)$ are 5.7 and 242.9 J K$^{-1}$ mol$^{-1}$, determine $\Delta_f S^\circ(CH_3NH_2, g)$ and $\Delta_f G^\circ(CH_3NH_2, g)$ at 298 K.

4.    (a) Write an equation for the formation of 1,1-dichloroethene from its constituent elements in their standard states. (b) Determine $\Delta S^\circ(298 \text{ K})$ given that $S^\circ(C, \text{graphite}) = 5.7$ J K$^{-1}$ mol$^{-1}$. (c) Is the reaction entropically favoured? (d) Look up the values of $\Delta_f H^\circ(298 \text{ K})$ and $\Delta_f G^\circ(298 \text{ K})$ for 1,1-dichloroethene. What conclusions do you draw from these values? (e) What can you say about the role that entropy plays in dictating whether the reaction in part (a) is thermodynamically feasible?

5.    At 298 K, $\Delta_f G^\circ(NH_3, g) = -16.4$ kJ mol$^{-1}$ but at 500 K, the value is $+4.8$ kJ mol$^{-1}$. Values of $\Delta_f H^\circ(NH_3, g)$ at 298 and 500 K are $-45.9$ and $-49.9$ kJ mol$^{-1}$ respectively. (a) What do the values of $\Delta_f H^\circ(NH_3, g)$ suggest about the formation of ammonia at 298 and 500 K? (b) Do the values of $\Delta_f G^\circ$ support these conclusions? (c) Determine $\Delta_f S^\circ(NH_3, g)$ at each temperature; comment on the relative magnitudes of the values, and of the $T\Delta S$ terms.

## ELECTROCHEMISTRY

### Cell diagrams and $E°_{cell}$

**Problem set 12.7**

**Standard reduction potentials required for this section are listed in Appendix 12 of H&C**

1.  Write down the spontaneous cell reaction described by each of the following cell diagrams:

    (a)  $Fe(s) \,|\, Fe^{2+}(aq) \,\vdots\, 2H^+(aq) \,||\, [H_2(g, \text{ 1 bar})] \, Pt$

    (b)  $Zn(s) \,|\, Zn^{2+}(aq) \,\vdots\, Ag^+(aq) \,|\, Ag(s)$

    (c)  $Pt \,|\, 2I^-(aq), I_2(aq) \,\vdots\, Ce^{4+}(aq), Ce^{3+}(aq) \,|\, Pt$

➤ **In a cell diagram, the reduction process appears on the right-hand side**

2.  For each cell in question 1, determine $E°_{cell}$.

3.  The value of $E°_{cell}$ for the cell shown below is 0.26 V.

    $$Pt \,|\, V^{2+}(aq), V^{3+}(aq) \,\vdots\, 2H^+(aq) \,||\, [H_2(g, \text{ 1 bar})] \, Pt$$

    (a) Write down the half-equations for the reduction and oxidation processes.
    (b) What is the spontaneous cell reaction? (c) Calculate $E°$ for the half-cell involving the vanadium ions.

➤ **See Figure 12.15 in H&C for an example of the cell design**

4.  A half-cell containing a zinc metal strip dipping into an aqueous solution of zinc nitrate is connected by a salt-bridge and a metal wire to a half-cell composed of a lead strip dipping into an aqueous solution of lead(II) nitrate.
    (a) Write down the half-equation for the spontaneous *reduction process*.
    (b) Write down a cell diagram to represent the spontaneous cell reaction.
    (c) What is the value of $E°_{cell}$?

### Using tables of standard reduction potentials

In this section, we practice using standard reduction potentials to calculate values of $E°_{cell}$ and predict the thermodynamic viability of reactions by applying equation 12.14.

➤ **See worked example 12.10 in H&C**

$$\Delta G°(298 \text{ K}) = z \times F \times E°_{cell} \qquad \Delta G° \text{ is in J mol}^{-1} \qquad (12.14)$$

where:   $E°_{cell}$ is in V
$z$ = number of moles of electrons transferred per mole of reaction
$F$ = Faraday constant = 96 500 C mol$^{-1}$

**Problem set 12.8**

1.  Write balanced equations for the spontaneous reactions which occur when the following pairs of half-cells are combined:

    (a) $Zn^{2+}(aq) + 2e^- \rightleftharpoons Zn(s)$   and   $Cl_2(aq) + 2e^- \rightleftharpoons 2Cl^-(aq)$

    (b) $2H^+(aq) + 2e^- \rightleftharpoons H_2(g)$   and   $Mg^{2+}(aq) + 2e^- \rightleftharpoons Mg(s)$

    (c) $Cl_2(aq) + 2e^- \rightleftharpoons 2Cl^-(aq)$   and   $Br_2(aq) + 2e^- \rightleftharpoons 2Br^-(aq)$

    (d) $H_2O_2(aq) + 2H^+(aq) + 2e^- \rightleftharpoons 2H_2O(l)$   and   $I_2(aq) + 2e^- \rightleftharpoons 2I^-(aq)$

    (e) $[IO_3]^-(aq) + 6H^+(aq) + 6e^- \rightleftharpoons I^-(aq) + 3H_2O(l)$
    and   $I_2(aq) + 2e^- \rightleftharpoons 2I^-(aq)$

2.    For each reaction in question 1, determine (a) $E°_{cell}$ and (b) $\Delta G°(298 \text{ K})$.
3.    In each overall reaction in question 1, which species is the oxidizing agent?
4.    Consider the two reduction half-equations:

$$O_2(g) + 2H^+(aq) + 2e^- \rightleftharpoons H_2O_2(aq) \qquad\qquad E° = +0.70 \text{ V}$$

$$H_2O_2(aq) + 2H^+(aq) + 2e^- \rightleftharpoons 2H_2O(l) \qquad\qquad E° = +1.78 \text{ V}$$

(a) If aqueous hydrogen peroxide were combined with aqueous $Fe^{2+}$ in the presence of acid, would $H_2O_2$ act as a reducing agent or an oxidizing agent? (b) Write an equation for the overall reaction. (c) If acidified aqueous $H_2O_2$ were combined with aqueous $[MnO_4]^-$, would $H_2O_2$ act as a reducing agent or an oxidizing agent? (d) Give the equation for the overall reaction.

➤ **The Nernst equation**

**See pages 610-612 in H&C for the Nernst equation and examples of its application**

Some of the applications of the Nernst equation are to determine :
  • $E_{cell}$ under non-standard conditions,
  • the dependence of a half-cell reduction potential on solution concentration,
  • the dependence of a half-cell reduction potential on pH if the half-cell contains $H^+$ or $[OH]^-$.

**Problem set 12.9**

**Molar gas constant, $R = 8.314 \text{ J mol}^{-1} \text{ K}^{-1}$**

1.    Find the value of $E(298 \text{ K})$ for the $Cu^{2+}/Cu$ couple when $[Cu^{2+}] = 0.05 \text{ M}$.
2.    If the concentration of $Ag^+(aq)$ in an $Ag^+/Ag$ half-cell is decreased from 1 M to 0.1 M at 298 K, by how much does the reduction potential change?
3.    In order that the $Sn^{2+}/Sn$ couple has a value of $E = -0.16 \text{ V}$ at 298 K, what concentration of $Sn^{2+}(aq)$ ions should be used in the half-cell?
4.    The half-cell:    $PbSO_4(s) + 2e^- \rightleftharpoons Pb(s) + [SO_4]^{2-}(aq)$
      is employed in lead storage batteries. If the sulfate ion concentration drops from 1 M to 0.4 M (298 K), by how much does the half-cell potential change?
5.    If the half-cell:    $[Cr_2O_7]^{2-}(aq) + 14H^+(aq) + 6e^- \rightleftharpoons 2Cr^{3+}(aq) + 7H_2O(l)$
      is prepared in aqueous solution at pH 5.0 at 298 K, what is the value of $E$ if the concentrations of $[Cr_2O_7]^{2-}$ and $Cr^{3+}$ are 0.5 and 0.005 M respectively?

## 12.6    SOLUBILITY PRODUCT CONSTANTS

In this section we consider *sparingly soluble* salts. You should be aware of the differences between the terms soluble, sparingly soluble, and insoluble.

**Problem set 12.10**

1.    Write expressions for $K_{sp}$ of the following salts: (a) barium sulfate; (b) magnesium phosphate; (c) iron(III) hydroxide; (d) lead(II) iodide.

➤ *Question:* Why would a calculation about $K_{sp}$ never involve a nitrate?

2.    The value of $K_{sp}(298 \text{ K})$ for barium fluoride is $1.84 \times 10^{-7} \text{ mol}^3 \text{ dm}^{-9}$. What is the solubility of this salt at 298 K?
3.    What is the solubility of $Cd(OH)_2$ at 298 K, if $\log K_{sp}$ is $-14.28$?
4.    The solubility of silver iodide in water at 298 K is $9.2 \times 10^{-9} \text{ mol dm}^{-3}$. Determine the value of $K_{sp}$.
5.    The solubility of silver(I) bromide is $1.37 \times 10^{-5} \text{ g}$ per 100 cm³ of water at 298 K. What is the value of $K_{sp}$ for this salt? [$A_r$ Ag = 108; Br = 80]

# 13 *p*-Block and high oxidation state *d*-block elements

> **Topics**
> - Structural and bonding aspects of species containing *p*-block atoms
> - Reactivity patterns of species containing *p*-block atoms
> - Thermodynamics: applications to some equilibria

## STRUCTURAL AND BONDING ASPECTS OF SPECIES CONTAINING *p*-BLOCK ATOMS

This section deals with the structures of some of the species formed by the *p*-block elements. The answers to the questions in problem set 13.1 lie within the main text of H&C. Rather than follow the elements by periodic group as we did in the main text, the workbook covers the compounds in an integrated way. This should encourage you to think more carefully about the reasons why particular structures are adopted. Information about hydrides of the *p*-block elements can be found in Chapter 11 of H&C.

### Structure and bonding

**Problem set 13.1**

1. Draw the structures of $B_2H_6$ and $Al_2Cl_6$ and compare the bonding in these molecules.
2. Draw the structures of $PF_3$ and $BF_3$. Why are they not structurally similar?
3. Write down the formula of boric acid. In the solid state, boric acid forms a layer structure. What makes the assembly of the layers possible?
4. Structure **13.1** gives a representation of the structure of $[B_5O_6(OH)_4]^-$. (a) What is the geometry at each type of boron centre? (b) Which B–O bonds possess $\pi$-character? How does the $\pi$-bonding arise?
5. (a) In a silicate, what is the characteristic geometry at each silicon centre? (b) A zeolite is an aluminosilicate. What are the structural features of zeolites that make them useful as (a) drying agents (molecular sieves), (b) catalysts, and (c) ion-exchange materials?
6. Draw the structures of the following anions: (a) $[SnCl_5]^-$; (b) $[SnCl_6]^{2-}$; (c) $[AlCl_4]^-$; (d) $[PF_6]^-$; (e) $[SbF_4]^-$; (f) $[SbF_5]^{2-}$.
7. Draw the structures of the following oxides of nitrogen, giving resonance structures where appropriate, and state whether each compound is paramagnetic or diamagnetic: (a) $N_2O$; (b) NO; (c) $NO_2$; (d) $N_2O_3$.
8. Use VSEPR theory to predict the structures of $XeF_2$, $XeF_4$ and $XeF_6$. Are the experimentally observed structures consistent with your predictions?

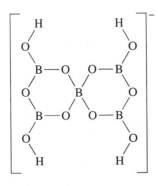

**(13.1)**

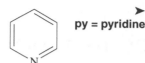

**py = pyridine**

9. What geometrical change takes place at the germanium centre when $GeCl_4$ forms the adduct *trans*-$[GeCl_4(py)_2]$?

10. What are the structural features that characterize an *oxoacid* ?

11. With the aid of diagrams, show the structural relationships between $P_4$, $P_4O_6$ and $P_4O_{10}$.

12. What structural feature of $H_3PO_3$ renders it dibasic rather than tribasic? On the other hand, why is $H_3PO_4$ tribasic?

13. (a) Explain why $N_2F_2$ has geometrical isomers. (b) How do the two structural isomers of $S_2F_2$ arise?

14. Compare the gas phase structures of $B_2Cl_4$ and $N_2F_4$ and account for any differences.

➤ **Azide anion, $[N_3]^-$: the answer is in Figure 5.2, H&C**

15. What shapes are (a) $O_3$, (b) $[I_3]^-$, and (c) $[N_3]^-$ ? Rationalize the shapes you have stated. Draw resonance structures to describe the bonding in $O_3$.

16. Which of the molecules $SO_2$, $SCl_2$, $Cl_2O$ and $ClO_2$ is/are paramagnetic? Rationalize your answer.

17. $N_2F_4$ exists in both staggered and *gauche* conformations, while the *gauche* conformation is favoured for $H_2O_2$ and the staggered conformation is favoured for $P_2I_4$. Draw the structures of these molecules in their respective conformations, and comment on factors that contribute towards the preference for a particular conformation.

➤ **Refer also to Chapters 3 and 7 in H&C.**

18. In the elemental state, oxygen occurs as diatomic molecules while sulfur occurs as $S_n$ rings or $S_\infty$ chains. Explain (a) how these different structural types are consistent with both elements being in group 16, and (b) why it is oxygen and not sulfur that preferentially forms diatomic molecules.

## Advanced study questions: NMR spectroscopy

**Problem set 13.2**

1. Predict the $^{19}F$ NMR spectrum of $[AsF_6]^-$.

2. Rationalize why the $^{31}P$ NMR spectrum of $[PF_6]^-$ is a septet.

3. Draw the structure of $[Sb_2F_{11}]^-$. Assuming free rotation around the Sb–F bonds, how many different $^{19}F$ environments are there in the anion?

**Data needed:**
$^{31}P$ 100%, $I = 1/2$
$^{19}F$ 100%, $I = 1/2$
**Assume that other nuclei are NMR *inactive***

4. Rationalize why the $^{19}F$ NMR spectrum of $PF_5$ consists of one signal which is a doublet at 298 K. [See p. 88 of the workbook for the $^{31}P$ NMR spectrum of $PF_5$.]

## REACTIVITY PATTERNS OF SPECIES CONTAINING *p*-BLOCK ATOMS

As in problem set 13.1, the answers to the problems below are found in Chapter 13 of the main text.

**Problem set 13.3**

1. Write equations to show how each of the following compounds behaves as an acid in the presence of water: (a) $B(OH)_3$; (b) $H_2SO_4$; (c) HBr; (d) $SO_2$; (e) $H_3PO_2$; (f) $H_3PO_4$; (g) $NO_2$.

2. What does *amphoteric* mean? Give equations to illustrate the amphoteric behaviour of aluminium oxide.

3. Explain what is meant by a Lewis acid-Lewis base adduct using the reaction of $BF_3$ and pyridine to illustrate your answer. Describe the changes in structure and bonding that occur at the boron centre in $BF_3$ during the reaction.

4. $AsF_5$ and $SbF_5$ are fluoride acceptors. Suggest products in the following reactions which are *not* necessarily balanced on the left-hand side:

(a) $CsF + AsF_5 \rightarrow$

(b) $HF + SbF_5 \rightarrow$

(c) $FNO + AsF_5 \rightarrow$

(d) $SbF_5 + [SbF_6]^- \rightarrow$

➤ **Refer also the Chapter 11 in H&C**

5. 'Aluminium and beryllium are *diagonally related* in the periodic table.' Give examples from the chemistries of these elements which support this statement. Name two other elements that are diagonally related.

6. The $pK_a$ of $[Ga(H_2O)_6]^{3+}$ is 2.6. Write an equation to show how this cation functions as an acid in aqueous solution, and explain *why* it is able to do so.

7. In small scale laboratory syntheses, which of the following gases could be collected by the displacement of water? (a) $H_2$; (b) $O_2$; (c) $SO_2$; (d) HCl; (e) $NH_3$; (f) $N_2$.

➤ **Refer also to Chapter 6 in H&C**

8. (a) Tin(IV) oxide has a rutile-type lattice. Draw a unit cell of the lattice and confirm the stoichiometry of the compound as $SnO_2$. (b) When $SnO_2$ reacts with aqueous HCl, in what form is the tin(IV) centre present? Draw the structure of the tin-containing species.

9. Give representative reactions that illustrate the Lewis acidity of (a) $GeCl_4$, (b) $GaBr_3$, (c) $SnF_4$ and (d) $AlCl_3$.

10. (a) Dioxygen can be prepared by the electrolysis of water. At which electrode is $O_2$ evolved? Write a half-equation to show its formation. (b) Rationalize the observed paramagnetism of $O_2$. (c) Under appropriate conditions (which you should specify), how does $O_2$ react with (a) Al; (b) amorphous carbon; (c) PbS; (d) $P_4$; (e) NO; (f) As; (g) $S_8$. [Remember: the conditions you specify may influence the products.]

11. Xenon(II) fluoride is commercially available. Give at least two reactions that demonstrate its role as a combined oxidizing *and* fluorinating agent.

12. Give two reactions in which $P_4O_{10}$ acts as a dehydrating agent; at least one should show how $P_4O_{10}$ may be used to form the acid anhydride of an inorganic acid.

13. Although the following equation is commonly seen in textbooks, there is, strictly, one error in it. What is it and how should the error be corrected?

$$I_2(aq) + 2Na_2S_2O_3(aq) \rightarrow 2NaI(aq) + Na_2S_4O_6(aq)$$

➤ ***Question:* Why should you never add water *to* concentrated sulfuric acid?**

14. How do the following elements or compounds react when *concentrated* $H_2SO_4$ is added to them? (a) calcium phosphate; (b) concentrated nitric acid; (c) copper; (d) sucrose; (e) water; (f) potassium chlorate (*caution!*). What application does reaction (b) have in organic chemistry?

15. 'Pure liquid $N_2O_4$ is *self-ionizing*.' What is meant by this statement and what applications does this liquid have in the laboratory?

➤
**Refer to Appendix 12 in H&C for clues**

16. Complete the following redox reactions which are *not* balanced on the left-hand side:

(a) $[ClO_3]^-(aq) + Fe^{2+}(aq) + H^+(aq) \rightarrow$

(b) $HNO_2(aq) \xrightarrow{\text{disproportionation}}$

(c) $Cu(s) + HNO_3(aq, \text{dilute}) \rightarrow$

(d) $I^-(aq) + Cl_2(g) \rightarrow$

(e) $[IO_3]^-(aq) + [SO_3]^{2-}(aq) \rightarrow$

---

## THERMODYNAMICS: APPLICATIONS TO SOME EQUILIBRIA

In this section we apply some of the equations and concepts from Chapter 12 to three equilibria. Two of the examples have been selected because of their relationship to environmental issues; emissions of $SO_2$ and $NO_x$ into the atmosphere may result in the formation of acid rain.

➤
**Acid rain: see Box 13.14 in H&C**

Answers are not provided for the case studies below; think about your answers and consider if the numerical answers you obtain are *sensible*.

### Case study 1: nitrogen oxide formation

The emissions from motor vehicles and aircraft are a major source of atmospheric pollution; in Box 13.9 in H&C, we looked at the role of catalytic converters in decreasing such emissions. One problem area is the formation of nitrogen oxides ($NO_x$) which occurs when atmospheric $N_2$ is oxidized by $O_2$ at the elevated temperatures reached when fuels are combusted in aviation and motor engines. We can represent this oxidation by equilibrium 13.1 which shows the formation of nitrogen oxide. Other oxides present in $NO_x$ mixtures include $NO_2$.

$$\tfrac{1}{2}N_2(g) + \tfrac{1}{2}O_2(g) \rightleftharpoons NO(g) \tag{13.1}$$

**13.1** A graph showing the temperature dependence of $\ln K_p$ for equilibrium 13.1.

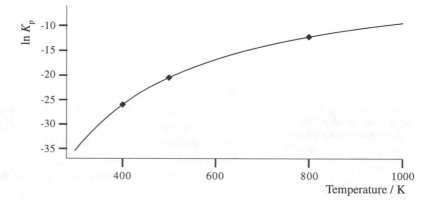

Figure 13.1 shows the temperature dependence of the equilibrium constant. At 298 K, this equilibrium lies well towards the left hand side:

$$\ln K_p(298 \text{ K}) = -35.3 \qquad\qquad K_p = 4.67 \times 10^{-16}$$

Calculate the value of $\Delta_f G^\circ(\text{NO, g, 298 K})$. Is the formation of nitrogen oxide thermodynamically favourable at room temperature?

A combustion engine operates at relatively high temperatures. From Figure 13.1, read off values of $\ln K_p$ at the points indicated and fill in the data below:

$$\ln K_p(400 \text{ K}) = \qquad\qquad\qquad K_p =$$

$$\ln K_p(500 \text{ K}) = \qquad\qquad\qquad K_p =$$

$$\ln K_p(800 \text{ K}) = \qquad\qquad\qquad K_p =$$

What do these values tell you about the position of equilibrium as a function of temperature?

Even at 800 K, the quantities of NO formed are small relative to the amount of $N_2$ and $O_2$ present in the atmosphere but are sufficient to pose a pollution threat.

### Case study 2: sulfur dioxide formation

Volcanic eruptions produce $SO_2$ naturally, but the combustion of fossil fuels which contain organo-sulfur compounds also results in significant emissions of $SO_2$ into the atmosphere. A relatively new fuel which contains smaller amounts of sulfur is *biodiesel*. As an exercise we consider the combustion of elemental sulfur, and Figure 13.2 shows how $\ln K_p$ for equilibrium 13.2 varies with temperature.

**Biodiesel: see Box 17.5 in H&C**

$$S(s) + O_2(g) \rightleftharpoons SO_2(g) \qquad\qquad (13.2)$$

**Activity: see Section 12.7 in H&C**

The *activity* of a pure solid is 1 and so the expression for $K_p$ involves only the partial pressures (strictly the activities) of $O_2$ and $SO_2$. Write down an equation for $K_p$. Use the data in Figure 13.2 to answer the following questions.

1. What are the values of $K_p$ at 300 and 600 K?
2. What do the values determined above tell you about the position of equilibrium at 300 and 600 K?
3. What are the values of $\Delta_f G(\text{SO}_2, \text{g})$ at 300 and 600 K?

**13.2** A graph showing the temperature dependence of $\ln K_p$ for the formation of sulfur dioxide from sulfur and dioxygen.

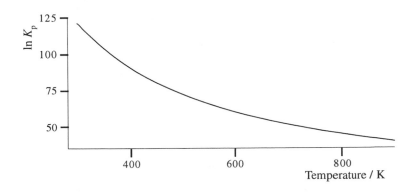

## Case study 3: dimerization of NO$_2$

➤

See Section 13.8 in H&C

Nitrogen dioxide can be prepared by heating lead(II) nitrate and if the stream of gaseous products is passed through a U-tube contained within a beaker of ice and water, yellow liquid N$_2$O$_4$ (bp 294 K) condenses in the tube. At 298 K the dimerization is represented by equation 13.3.

$$2NO_2(g) \rightleftharpoons N_2O_4(g) \qquad\qquad K_p(298\ K) = 8.7 \qquad\qquad (13.3)$$

The following questions relate to this equilibrium.

1.    Determine $\Delta_r G(298\ K)$ for the dimerization of NO$_2$(g).
2.    Is the dimerization of NO$_2$ entropically favoured? Rationalize your answer.
3.    How would the magnetic properties of a mixture of NO$_2$ and N$_2$O$_4$ assist in determining the composition of the mixture?

Figure 13.3 shows the variation of ln $K_p$ for the gas phase dissociation of N$_2$O$_4$ (equation 13.4).

$$N_2O_4(g) \rightleftharpoons 2NO_2(g) \qquad\qquad\qquad\qquad (13.4)$$

1.    Write an expression for $K_p$ for equilibrium 13.4. How is this expression related to that for $K_p$ for the gas phase dimerization of NO$_2$?
2.    Using Figure 13.3, complete the following data:

ln $K_p$(350 K) =                    $K_p$(350 K) =

ln $K_p$(500 K) =                    $K_p$(500 K) =

3.    What do the values you have determined tell you about the position of equilibrium as the temperature increases?
4.    Determine a value of $\Delta_r G$ at each temperature.
5.    If $\Delta_r H(350\ K) = 58$ kJ per mole of N$_2$O$_4$, calculate $\Delta_r S(350\ K)$. Evaluate your answer in terms of the dissociation process, paying particular attention to the sign of $\Delta S$.

**13.3**  A graph showing the variation in ln $K_p$ over the temperature range 298 to 600 K for the dissociation of N$_2$O$_4$.

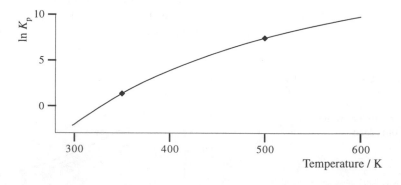

# 14  Polar organic molecules

---

**Topics**

- Polar bonds and molecular dipole moments
- Halogenoalkanes
- Ethers, alcohols and amines
- Multi-step syntheses

---

## POLAR BONDS AND MOLECULAR DIPOLE MOMENTS

### Molecular dipole moments: $CH_3F$, $CH_2F_2$ and $CHF_3$

➤
**Pauling electronegativity values:**

$\chi^P(C) = 2.6$

$\chi^P(H) = 2.2$

$\chi^P(F) = 4.0$

In Section 14.2 of H&C, we stated that it can be dangerous to draw rapid conclusions about the magnitude and direction of a molecular dipole moment. In order to explore this point further, we consider a series of fluoroalkanes. The experimental *gas phase* values for the molecular dipole moments of $CH_3F$, $CH_2F_2$ and $CHF_3$ are 1.86, 1.98 and 1.65 D respectively. At first this may seen strange — should the dipole moment not increase as the number of fluorine atoms increases? Each of the molecules is tetrahedral with bond angles of *approximately* 109.5°. Figure 14.1 shows the structures of the three molecules and, by using Pauling electronegativity values, we can suggest in which direction the molecular dipole acts in each molecule.

Now let us quantify this picture. An *approximate* method of predicting the molecular dipole moments is to estimate the *bond* dipole moments using Pauling electronegativity values, and then resolve the bond moments into a single direction.

➤
**Resolution of bond dipole moments: see Box 5.1 in H&C**

*Approximate* values for the C–H and C–F bond dipole moments (in debye) are estimated below:

Bond dipole moment for C–H $\approx \chi^P(C) - \chi^P(H) = 2.6 - 2.2 = 0.4$ D $\qquad C—H$

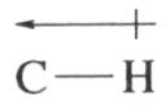

Bond dipole moment for C–F $\approx \chi^P(F) - \chi^P(C) = 4.0 - 2.6 = 1.4$ D $\qquad C—F$

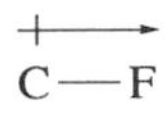

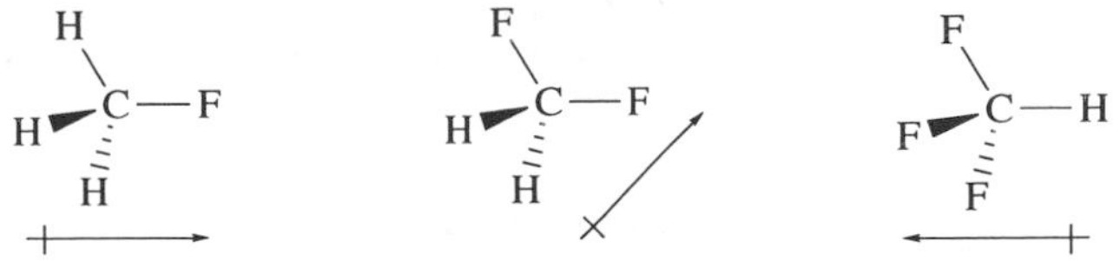

**14.1** By considering Pauling electronegativity values, the direction in which the molecular dipole acts in each of $CH_3F$, $CH_2F_2$ and $CHF_3$ can be deduced.

The molecular dipole moment in $CH_2F_2$ is now estimated as follows:

Direction along which bond dipole moments are resolved.

Angle $\alpha = \dfrac{109.5}{2} = 54.75°$

$\left.\begin{array}{l}\text{The resultant dipole moment}\\\text{due to the (C – F) contributions}\end{array}\right\} = 2 \times (1.4 \times \cos 54.75) = 1.62 \text{ D}$

$\left.\begin{array}{l}\text{The resultant dipole moment}\\\text{due to the (C – H) contributions}\end{array}\right\} = 2 \times (0.4 \times \cos 54.75) = 1.46 \text{ D}$

These both act in the same direction and the total molecular dipole moment for $CH_2F_2$ is estimated to be 2.08 D.

The molecular dipole moment in $CHF_3$ is estimated as follows:

Direction along which bond dipole moments are resolved.

Angle $\alpha = 180 - 109.5 = 70.5°$

$\left.\begin{array}{l}\text{The resultant dipole moment}\\\text{due to the (C – F) contributions}\end{array}\right\} = 3 \times (1.4 \times \cos 70.5) = 1.4 \text{ D}$

$\left.\begin{array}{l}\text{The resultant dipole moment}\\\text{due to the (C – H) contributions}\end{array}\right\} = 0.4 \text{ D}$

These both act in the same direction and the total molecular dipole moment for $CHF_3$ is estimated to be 1.8 D. As an exercise, you should show that $CH_3F$ has an estimated dipole moment of 1.8 D, acting in the direction shown in Figure 14.1.

Although the estimated values for the three fluoroalkanes do not agree exactly with the experimental values, we can at least begin to understand why the molecular dipole moment does *not* simply follow the order $CHF_3 > CH_2F_2 > CH_3F$. However, a *word of caution*: this approach does not allow you to rationalize all the experimentally observed trends in molecular dipole moments.

## Polar molecules

*Exercise*: Which of the following molecules are polar? $CH_3Cl$; $CH_3OH$; $CH_2Br_2$; $H_2C=O$; $CO_2$; $H_2O$; HBr; $Me_2C=O$; $CH_3CH_2OH$; $CH_3CH_2F$; $Me_3COH$; $CBr_4$.

## HALOGENOALKANES

➤
**Before working through
the next set of problems,
you should have studied
Sections 14.4–14.6 in H&C**

Some of the problems below illustrate the competition between nucleophilic substitution and β-elimination reactions of halogenoalkanes. In other cases, for the sake of simplicity, we assume that a particular pathway predominates; consider *to what extent* and *under what conditions* the assumptions may be valid.

**Problem set 14.1**

1. Suggest methods of preparing (a) 2-chlorobutane, (b) 1,2-dibromopropane and (c) chloromethane.

2. Suggest products for the following reactions:

   (a) $CH_3CH_2Cl + 2Li \xrightarrow{\text{anhydrous Et}_2O}$

   (b) $CH_3CH_2Cl + CH_3CH_2CH_2CH_2Li \rightarrow$

   (c) $CH_3CH_2CH_2Cl + Mg \xrightarrow{\text{anhydrous Et}_2O}$

   (d) $CH_3CH_2Br + Na[CH_3CH_2O] \rightarrow$

3. Outline the mechanism of an $S_N1$ reaction, using as an example the reaction between water and 2-chloropropane. Include in your answer a rate equation for the reaction.

4. Outline the mechanism of an $S_N2$ reaction, using as an example the reaction between hydroxide ion and 1-chlorobutane. Include in your answer a rate equation for the reaction.

5. What do you understand by the terms *retention* and *inversion of configuration*?

6. If the reaction of hydroxide ion with a single enantiomer of MeEtBuCX leads to a racemic mixture of alcohols, what can you deduce about the mechanism of the reaction?

7. $S_N1$ and $S_N2$ mechanisms are also referred to as dissociative and exchange mechanisms respectively. Why are the latter names commonly used?

8. (a) If you were asked to prepare a solution of *alcoholic KOH*, what reagents would you use? (b) Does alcoholic KOH act as a strong or weak base? (c) What would you expect to be the major product of the reaction between alcoholic KOH and 1-bromopropane?

9. The treatment of one enantiomer of 2-iodooctane with Na(*I) [where *I is isotopically labelled iodine] leads to racemization. The rates of iodine-exchange and inversion are identical. (a) Write a mechanism consistent with these observations. (b) How is the rate of racemization related to the rate of exchange?

10. Experimental data show that the reaction of a particular halogenoalkane $R_3CX$ with water follows first order kinetics overall. (a) Write down an expression for the rate equation. (b) Do these data *alone* allow you to distinguish between the reaction being a substitution, an elimination, or a competition between the two?

## ETHERS, ALCOHOLS AND AMINES

1. The $^{13}C$ NMR spectrum of a structural isomer of butanol has three signals ($\delta$ 18.9, 30.8 and 69.4) with relative integrals 2 : 1 : 1. (a) Draw the structure of the isomer. (b) What is its systematic name?

2. Figure 14.2 shows the IR spectrum of hexadecan-1-ol. (a) What is the formula of hexadecan-1-ol? (b) Which features can you assign in the IR spectrum and which lie in the fingerprint region?

3. Figure 14.3 shows the IR spectrum of $(Me_2CHCH_2CH_2)_2O$. (a) To what do you assign the absorptions just below 3000 cm$^{-1}$? (b) Which absorption may be assigned to the stretching mode of the C–O–C group? (c) Why is there a broad absorption centred at $\approx$3300 cm$^{-1}$?

4. Outline the practical problems associated with the Williamson synthesis of an asymmetrical ether.

5. (a) Why is the boiling point of diethyl ether significantly lower than that of butan-1-ol even though the values of $M_r$ are the same? (b) Why must you take safety precautions when working with $Et_2O$ in the laboratory?

*Caution!* ➤

6. The $^{13}C$ NMR spectrum of 2,5-dioxahexane consists of two equal intensity signals at $\delta$ 58.6 and 72.3. Draw the structure of 2,5-dioxahexane and assign the spectrum.

**14.2** The IR spectrum of hexadecan-1-ol.

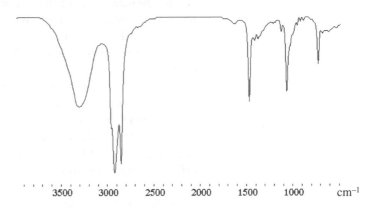

**14.3** The IR spectrum of $(Me_2CHCH_2CH_2)_2O$.

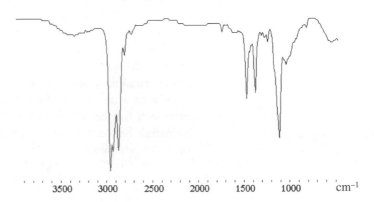

7. $CH_3SCH_2CH_3$ is a *thioether*, structurally related to an ether but with an $-S-$ unit replacing the ether functionality. The $^1H$ NMR spectrum of $CH_3SCH_2CH_3$ shows a singlet ($\delta 2.1$), a quartet ($\delta 2.5$) and a triplet ($\delta 1.3$). Assign the spectrum.

8. Suggest products for the following reactions:

   (a) $CH_3CH_2CH(OH)CH_3 \xrightarrow{Al_2O_3,\ heat}$

   (b) $CH_3CH_2CH_2OH + SOCl_2 \rightarrow$

   (c) $Me_3COH + Na \rightarrow$

   (d) $H_2NCH_2CH(NH_2)CH_3 \xrightarrow{excess\ HBr}$

   (e) $CH_3CH_2C\equiv N \xrightarrow{Li[AlH_4]}$

   (f) $Me_2CH(OH) \xrightarrow{acidified\ K_2Cr_2O_7}$

9. Methods of preparing alcohols include the hydroboration of alkenes and the acid catalysed addition of water to alkenes. Starting from but-1-ene, show how and why the products of these methods of synthesis may differ.

10. What would be the most diagnostic features that would differentiate the IR spectra of ethylamine and triethylamine?

➤ **See Appendix 11 in H&C for data needed**

11. When 2.60 g of dibutyl ether are fully combusted, 106.9 kJ of heat energy (measured at 298 K) are given out to the surroundings. Determine a value for $\Delta_f H^o$(dibutyl ether, l, 298 K). [$A_r$ C = 12, O = 16, H = 1]

12. Suggest possible products for the following reactions:

   (a) $CH_3CH_2OCH_2CH_3 \xrightarrow{concentrated\ HBr}$

   (b) $CH_3CMe_2CMe_2OH \xrightarrow{PCl_3}$

   (c) $Me_3COH \xrightarrow{acidified\ K_2Cr_2O_7}$

   (d) $CH_3CH_2OH \xrightarrow{Corey's\ reagent}$

   (e) $HOCH_2CH_2OH \xrightarrow{acidified\ K_2Cr_2O_7}$

   (f) $[Et_4N]OH + CH_3CO_2H \rightarrow$

---

## MULTI-STEP SYNTHESES

Although organic reactions are often considered individually in lecture courses and textbooks, you are most likely to prepare compounds in the laboratory by combining reactions into multi-step syntheses. Such strategies are fundamental to most industrial processes where saturated and unsaturated hydrocarbons are the basic chemicals from which others are derived; for reasons of practicality and economics, starting materials must obviously be readily available.

In this section, we combine some of the reactions of alkanes, alkenes and polar molecules to provide multi-step synthetic pathways. Since there may be more than one possible answer to each question, answers have not been provided in the workbook, but may be found by reading Chapters 8 and 14 of H&C.

**Problem set 14.3**    1.    Give suitable reaction conditions under which to carry out the following multi-step syntheses.

2.    How would you undertake the following transformations? You may use as many steps as you think are necessary, but side products should be minimized.

3.    (a) Suggest identities for the numbered compounds in the following reaction scheme. Begin with the alkene highlighted in the box.

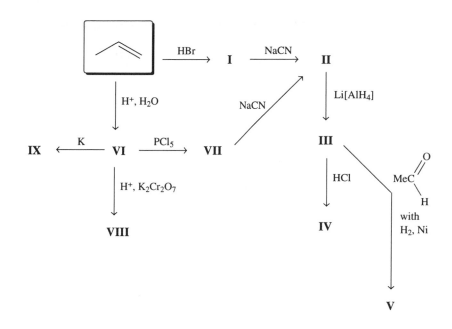

(b) Apply what you know about mass spectrometry and IR and NMR spectroscopies to indicate how the nature of the products might be established.

(c) To which numbered compound in the above scheme could the $^{13}C$ NMR spectrum shown below be assigned? (The effects of $^1H-^{13}C$ coupling have been removed from the spectrum.)

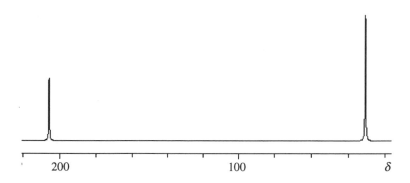

4.    Suggest suitable conditions which you could use to carry out the transformations shown in the following reaction scheme. Begin at the compound highlighted in the box. Point out any competing reactions that may occur.

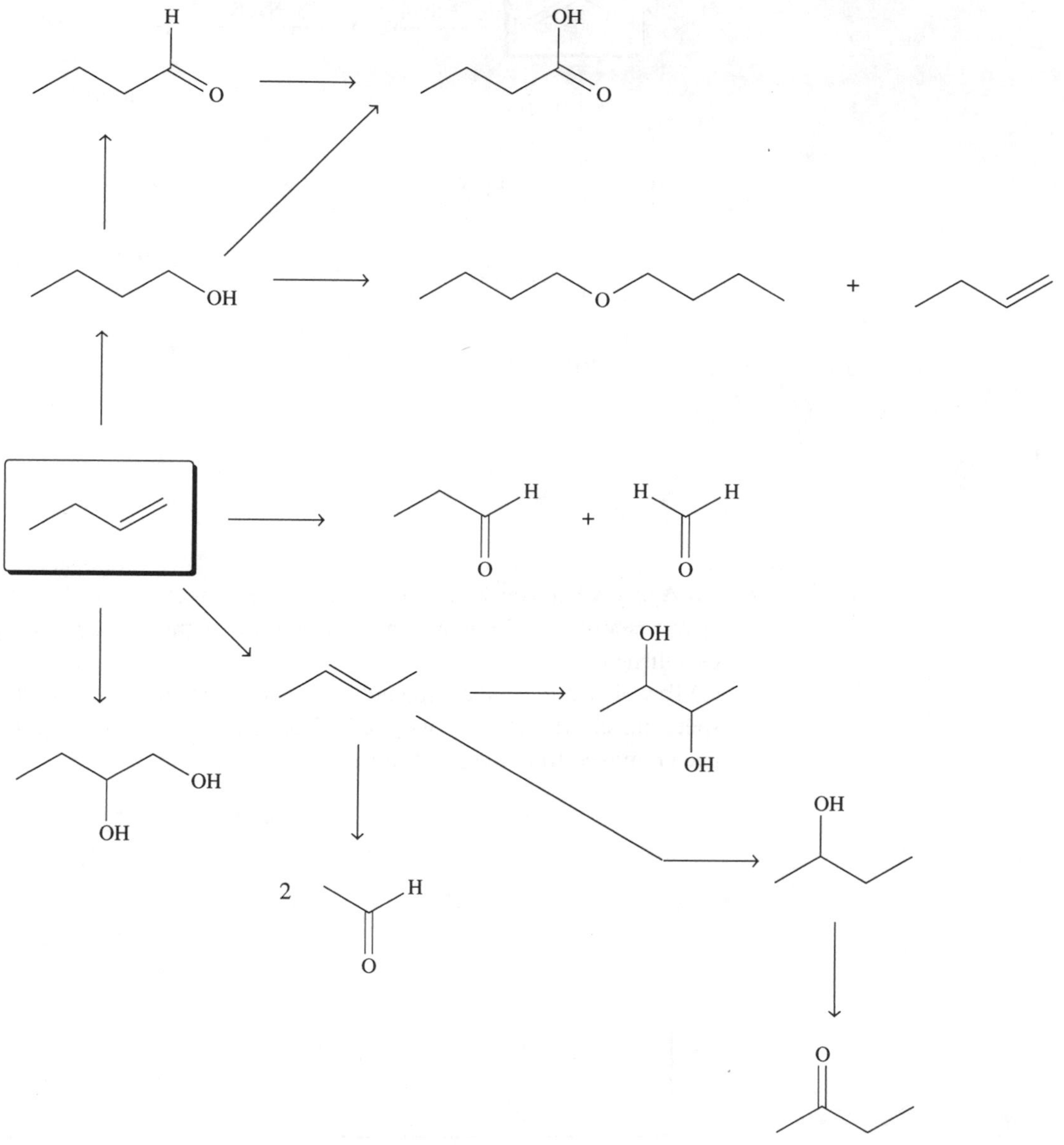

# 15 Rings

---

**Topics**
- Cycloalkanes, cycloalkenes and some derivatives
- Benzene and its derivatives
- Activation, deactivation and orientation effects

---

## CYCLOALKANES, CYCLOALKENES AND SOME DERIVATIVES

The questions in this section cover material from Sections 15.2 to 15.4 of H&C.

### Structure, physical properties and spectroscopy

**Problem set 15.1**

**15.1** The structure of cyclohexane.

1. Explain why cyclopropane is a relatively rigid molecule but cyclodecane can adopt numerous conformations.
2. Figure 15.1 shows the structure of $C_6H_{12}$. (a) Name the conformation illustrated. (b) Mark the axial and equatorial hydrogen positions.
3. Draw the structures of (a) cyclopentadiene, (b) cyclooctane, (c) ethylene oxide, (d) cyclohexa-1,3-diene and (e) 1,2-dibromocyclopropane (including isomers).
4. The IR spectrum and two views of the structure of menthol are shown in Figure 15.2. (a) How many asymmetric carbon atoms does menthol possess? (b) Which absorptions in the IR spectrum of menthol can you assign? (c) If the ring in Figure 15.2b 'flipped' into another chair-conformation, would the substituents still occupy equatorial sites?

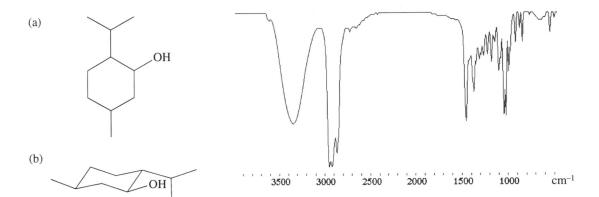

**15.2** The structure (two views) and IR spectrum of menthol. Menthol is used as an inhalant to relieve nasal congestion.

**15.3**   100 MHz $^{13}$C NMR
spectrum of cyclohexanone.

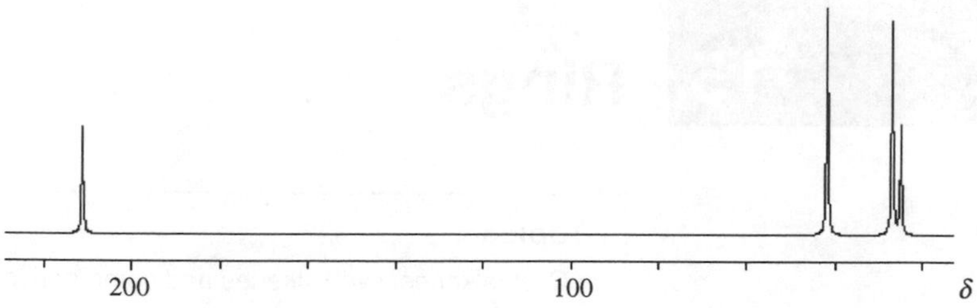

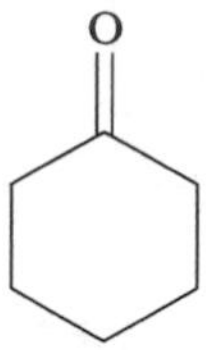

➤
**Refer to Table 9.11 in H&C**

5.   Figure 15.3 shows the $^{13}$C NMR spectrum of cyclohexanone. Which signals can you unambiguously assign?

6.   (a) THF is a common laboratory solvent; for what is THF an abbreviation? (b) Draw the structure of THF and show the direction in which the dipole moment acts. (c) The dipole moment of THF (in the gas phase) is 1.75 D, while that of 1,3-dioxolane is 2.06 D. Account for this difference.

## Synthesis and reactivity

**Problem set 15.2**

1.   Suggest products for the following reactions, commenting where the pathways may be non-specific.

(a) $CH_2$=$CHCH_3$ + $CH_2N_2$ $\xrightarrow{hv}$

(b) 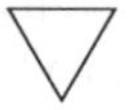 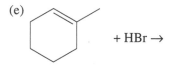 + $H_2$ $\xrightarrow{\Delta, \text{ Ni catalyst}}$

(c) 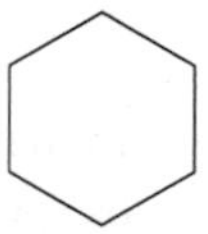   + $Cl_2$ $\xrightarrow{hv}$

(d) 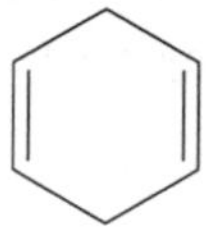   + $H_2$ $\xrightarrow{\Delta, \text{ Pd catalyst}}$

(e)   + HBr →

2.   Using selected reactions, show how the reactivity of cyclopropane is *not* typical of a saturated molecule. Suggest reasons for this behaviour.

3.   What is a *Diels-Alder* reaction? Show how this class of reaction can be used to prepare cycloalkenes.

4.   What products may be obtained when acetylene is heated in the presence of a nickel catalyst? Write a mechanism for each transformation that you describe.

5.   The reaction of cyclohexene with sulfuric acid and water gives compound

**15.4** For question 5, problem set 15.2.

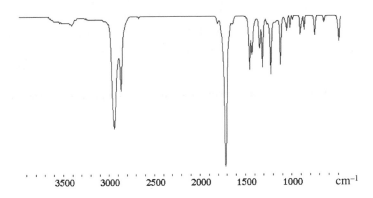

**A**. When **A** is treated with acidified potassium dichromate, the product is **B**, the IR spectrum of which is shown in Figure 15.4. (a) Suggest identities for **A** and **B**. (b) What features of the IR spectrum are you able to assign? (c) Write an equation for the complete combustion of compound **A**.

## BENZENE AND ITS DERIVATIVES

The problems in this section cover material from Sections 15.5–15.12 in H&C. This allows us to explore interconversions of various derivatives of benzene as well as the chemistry of $C_6H_6$.

### Structure, bonding and spectroscopy

Problem set 15.3

1. Outline the bonding theories for benzene that account for the fact that the C–C bonds are of equal length.

2. (a) Describe the bonding in the phenoxide ion both in terms of an MO picture and resonance structures. (b) How does the bonding picture you have described account for the acidity of phenol?

3. $^1$H NMR spectroscopic data for three aromatic compounds **A**, **B** and **C** are as follows:

| **A** | $\delta$ 7.2 (multiplet), 2.3 (singlet) | Rel. integrals 5 : 3 |
|---|---|---|
| **B** | $\delta$ 7.3 (multiplet), 4.5 (singlet), 2.4 (broad) | Rel. integrals 5 : 2 : 1 |
| **C** | $\delta$ 7.4–6.8 (multiplet), 3.8 (singlet) | Rel. integrals 5 : 3 |

One of the compounds ($M_r = 92$) contains 91.3% C and 8.7% H; the other two compounds (both $M_r = 108$) contain 77.8% C and 7.4% H. Suggest identities for **A**, **B** and **C** and assign the $^1$H NMR spectra.

4. How might $^{13}$C NMR spectroscopy be used to distinguish between 1,3,5-$Me_3C_6H_3$ and 1,2,4-$Me_3C_6H_3$?

5. The product of a particular synthesis is a chlorinated derivative of benzene $C_6H_{6-x}Cl_x$ which contains 48.3% Cl; two isomers have been separated. The

$^{13}C$ NMR spectrum of isomer **A** consists of three equal intensity signals while that of isomer **B** consists of two signals in a ratio 1 : 2. (a) What is the molecular formula of the derivative. (b) Identify isomers **A** and **B**. (c) What is the structure of the third possible isomer, and how would its $^{13}C$ NMR spectrum differ from those of **A** and **B**?

6.  Which of the following compounds are aromatic? Give reasons for your choices.

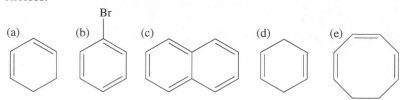

## Syntheses and reactivity

**Problem set 15.4**

1.  Suggest products for the reactions of benzene with:
    (a) Na in liquid $NH_3$ in the presence of $^tBuOH$;
    (b) $Cl_2$ in the presence of $AlCl_3$;
    (c) $CH_3Cl$ in the presence of $AlCl_3$, followed by treatment with hot alkaline $KMnO_4$;
    (d) a mixture of concentrated sulfuric and nitric acids.

2.  How might the following compounds be prepared *starting from benzene*?
    (a) phenol; (b) aniline; (c) sodium benzoate; (d) a mixture of 1,2- and 1,4-dimethylbenzene (1,2- and 1,4-xylene).

3.  Draw a scheme to show the mechanism and products of the reaction between benzene and $Br_2$ in the presence of the Lewis acid $FeBr_3$. State clearly which species are intermediates or transition states.

4.  Explain why the major product of the reaction of benzene with 1-chloropropane in the presence of $AlCl_3$ is $C_6H_5CHMe_2$ rather than $C_6H_5CH_2CH_2CH_3$.

5.  When toluene reacts with $Cl_2$, the products depend upon the conditions used in the experiment. Explain why this is, and give the conditions that you would use to form the two compounds shown in Figure 15.5. What side-products might be expected in each case?

6.  In what ways do the reactivity patterns of the following compounds differ, and why? Include in your answer selected reactions that highlight the behaviours of *both* compounds.

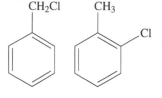

**15.5**  For question 5, problem set 15.4.

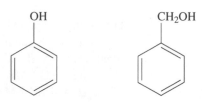

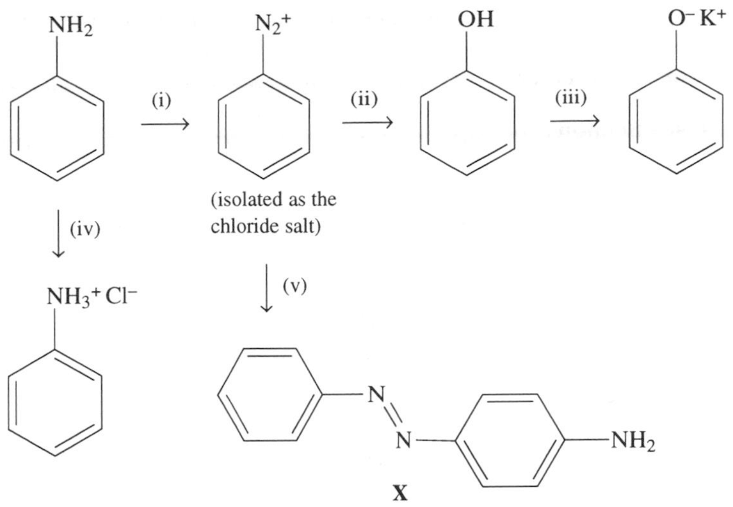

**15.6** Reaction scheme for question 6, problem set 15.4.

7.  (a) Give suitable reaction conditions [numbered (i) to (v)] for the transformations shown in Figure 15.6. (b) What type of compound is product **X** in Figure 15.6? (c) Identify the chromophore in compound **X**, and state what properties may be exhibited by compounds containing such a unit.

8.  Draw the structure of pyridine and write an equation to show how it acts as a Brønsted base in aqueous solution.

➤ 9.  Suggest products for the reactions of pyridine with (a) HBr(aq); (b) $CH_3CO_2H(aq)$; (c) $C_2H_5I$; (d) $HBF_4(aq)$; (e) $BF_3$.

**Tetrafluoroboric acid: see equation 13.7 in H&C**

## ACTIVATION, DEACTIVATION AND ORIENTATION EFFECTS

➤ **See Table 15.4 in H&C**

In Section 15.8 of H&C, we described the *ortho-*, *meta-* and *para-*directing properties of some substituent groups in monosubstituted aromatic rings, and whether the substituent activated or deactivated the ring towards further attack by an electrophile. We now consider the reasons why these orientation effects arise.

It is important to recognize that the observation that a particular substitution product is favoured, and that a ring is activated or deactivated are *kinetic effects*. For example, the presence of the methyl group in toluene activates the ring towards further attack by Me$^+$ and electrophilic substitution preferentially occurs at the 2- and 4-positions because reaction at these sites occurs more rapidly than substitution at the 3-position.

The rate determining step in an electrophilic substitution is often the formation of a Wheland intermediate. Consider reactions 15.1 and 15.2.

$$C_6H_6 + Y^+ \rightarrow C_6H_5Y + H^+ \tag{15.1}$$

$$C_6H_5X + Y^+ \rightarrow C_6H_4XY + H^+ \tag{15.2}$$

➤

**Stabilization of carbenium ions by electron releasing groups: see Section 8.14 in H&C**

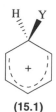

**(15.1)**

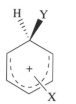

Substituent X is not in a particular position.

**(15.2)**

The mechanisms for these reactions involve Wheland intermediates **15.1** and **15.2** respectively. Compared to intermediate **15.1**, the positive charge in **15.2** is *stabilized* if *substituent X is electron releasing*. If in structure **15.2** substituent X is electron withdrawing, the carbenium ion is *destabilized*. Whether or not a monosubstituted derivative $C_6H_5X$ is activated or deactivated with respect to benzene depends on the electron releasing or withdrawing properties of group X. Look at Table 15.4 in H&C and correlate the activating or deactivating properties of the substituents listed with what you know about their electronic properties. For example, Cl is electronegative and so is electron withdrawing.

In order to understand the origins of *ortho-*, *meta-* and *para-*directing groups, we need to consider the set of resonance structures for the Wheland intermediate — these tell us the distribution of positive charge in the carbenium ion (see Figure 15.14 in H&C). Let us take the case of reaction 15.2 in which electrophile $Y^+$ attacks in the position *para* to group X. Figure 15.7a shows the resonance structures for the intermediate formed in the rate determining step of the reaction. Of these three carbenium ions, the middle one is particularly stable if X is *electron releasing*, but is destabilized if X is *electron withdrawing*. Figure 15.7b shows one of three resonance structures for the intermediate formed in the rate determining step of reaction 15.2 if $Y^+$ attacks the site *ortho* to group X. [*Exercise*: draw the other two resonance structures.] The resonance structure shown is particularly stable if X is *electron releasing*, but is destabilized if X is *electron withdrawing*. We may conclude that *ortho-* and *para-*substitutions are *both favoured* in reaction 15.2 if group X releases electrons to the ring. Since the positive charge cannot be localized on the *meta*-position, electron releasing groups are not *meta*-directing.

What happens if X is electron withdrawing? In Figure 15.7a, the middle resonance structure is particularly *unfavourable* if X withdraws electron from the carbenium centre, and similarly in Figure 15.7b. Hence, *ortho-* and *para*-attack are not favoured for reaction 15.2 if X is electron withdrawing, and as a result, group X is observed to be *meta*-directing.

If we look at Table 15.4 in H&C, the orientation effects of Cl and Br substituents do not appear to fit the pattern that we have just described. What causes these electron withdrawing groups to be *ortho-* and *para*-directing? The answer lies in the fact that the resonance structures shown in Figures 15.7c and 15.7d contribute significantly when X = Cl or Br and make *para-* and *ortho*-substitution favourable for reactions of chloro- or bromobenzene with electrophiles.

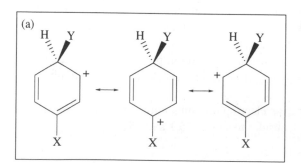

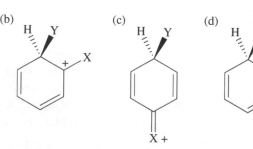

**15.7**    Resonance structures for the intermediates formed when $Y^+$ attacks the monosubstituted derivative $C_6H_5X$. See text for details.

# 16 Coordination complexes of the *d*-block metals

> **Topics**
> - Structure and isomerism
> - Hexaaqua ions and redox behaviour
> - Complex formation and stability constants
> - Crystal field theory, the spectrochemical series and magnetism
> - Metal carbonyl complexes

## STRUCTURE AND ISOMERISM

> **Ligand abbreviations used in this chapter are listed in Table 16.2 of H&C**

In this section, we focus on the structures of coordination complexes, and include practice in recognizing common ligands. You should be familiar with the following arrangements of donor atoms around a metal centre:
- linear
- trigonal planar
- tetrahedral
- square planar
- trigonal planar
- square-based pyramidal
- octahedral

**Problem set 16.1**

> **Complexes of platinum(II) and gold(III) are usually square planar**

1. Draw the structures of the following ligands showing whether they are charged or neutral, and indicating the donor atoms: (a) oxalate (ethanedioate); (b) 1,2-diaminoethane; (c) thiocyanate; (d) acetylacetonate; (e) pyridine; (f) 2,2'-bipyridine; (g) acetonitrile; (h) cyanide; (i) [edta]$^{4-}$. What are the common abbreviations for ligands (a), (b), (d), (e) and (f) ?

2. Give *two* examples of tridentate ligands.

3. Draw the structures of the following complexes, indicating the presence of isomers where appropriate:
   (a) [PtCl$_2$(NH$_3$)$_2$]; (b) [AuCl$_4$]$^-$; (c) [Zn(OH)$_4$]$^{2-}$; (d) [Fe(CN)$_6$]$^{4-}$; (e) [Ni(en)$_3$]$^{2+}$; (f) [Cr(NH$_3$)$_4$Cl$_2$]$^+$; (g) [V(ox)$_3$]$^{3-}$; (h) [Co(en)$_2$F$_2$]$^+$.

4. Figure 16.8 in H&C shows the structure of [Fe(tpy)$_2$]$^{2+}$. Explain why only the *mer-* (and not the *fac-*) form of this complex is known.

5. The compound [CrCl$_2$(H$_2$O)$_4$]Cl·2H$_2$O is commercially available. (a) What is the oxidation state of the chromium centre? (b) Draw the structures of the isomers of the complex cation present in this compound. (c) What other types of isomerism may [CrCl$_2$(H$_2$O)$_4$]Cl·2H$_2$O exhibit?

6. Which of the following complexes possess enantiomers? (a) [Cr(acac)$_3$]$^{3-}$; (b) [Cr(NH$_3$)$_2$(NCS)$_4$]$^-$; (c) [Cr(bpy)$_3$]$^{2+}$; (d) [Fe(bpy)$_2$Cl$_2$].

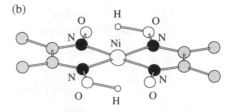

**16.1** The structure of the complex [CoCl₃(PEt₃)₂] determined by X-ray crystallography (Et = ethyl).

**16.2** The structure of (a) the ligand dimethylglyoxime, H₂dmg, and (b) the nickel(II) complex [Ni(Hdmg)₂].

7.   Figure 16.1 shows the solid state structure of the complex [CoCl₃(PEt₃)₂] (Et = ethyl). (a) What is the oxidation state of the cobalt centre? (b) Suggest reasons for the preferential siting of the chloride ligands in the equatorial plane.

8.   Figure 16.2a shows the structure of dimethylglyoxime, abbreviated to H₂dmg. This ligand is used for the quantitative analysis of nickel; it forms a deep red complex [Ni(Hdmg)₂] (Figure 16.2b). (a) Draw resonance structures for the ligand [Hdmg]⁻. (b) Nickel(II) forms both tetrahedral and square planar complexes. Why do you think that [Ni(Hdmg)₂] is planar?

## Chiral ligands

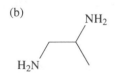

**16.3** The structures of (a) 1,3-diaminopropane and (b) 1,2-diaminopropane.

The complex *trans*-[Co(en)₂Cl₂]⁺ is *achiral*; if you draw the structure of this complex and that of its mirror image, the mirror image is superimposable upon the first structure. On the other hand, *cis*-[Co(en)₂Cl₂]⁺ is *chiral* and the chirality arises from the arrangement of the donor atoms around the metal centre.

Now consider the ligand 1,3-diaminopropane (abbreviated to 1,3-pn) (Figure 16.3a). The complex *trans*-[Co(1,3-pn)₂Cl₂]⁺ is achiral and the *cis*-isomer is chiral for the same reasons that *trans*-[Co(en)₂Cl₂]⁺ and *cis*-[Co(en)₂Cl₂]⁺ are optically inactive and optically active respectively.

Look next at the structure of 1,2-diaminopropane (1,2-pn) drawn in Figure 16.3b. This ligand possesses an asymmetric carbon centre — which atom is this in the figure? When 1,2-diaminopropane coordinates to a metal centre, the ligand retains its chiral centre and may impart chirality on the complex. To understand this, consider the complex *trans*-[Co(1,2-pn)₂Cl₂]⁺. The problem is not a simple one. Whereas there is only one structure for *trans*-[Co(1,3-pn)₂Cl₂]⁺ (assuming restricted rotation about the C–C and C–N bonds in the ligands), there are six structures

**16.4**  Schematic representation of one isomer of *trans*-[Co(1,2-pn)$_2$Cl$_2$]$^+$.

possible for *trans*-[Co(1,2-pn)$_2$Cl$_2$]$^+$, two pairs of enantiomers and two achiral forms. *Exercise*: draw these six isomers, one of which is shown in Figure 16.4.

## HEXAAQUA IONS AND REDOX CHEMISTRY

**Problem set 16.2**

1.  In aqueous solution, the p$K_a$ values of the hexaaqua ions [Ti(H$_2$O)$_6$]$^{3+}$ and [Fe(H$_2$O)$_6$]$^{3+}$ are 3.9 and 2.0 respectively. (a) Determine values of $K_a$ for the two complex ions. (b) Write equations for the equilibria to which these constants refer. (c) Which complex ion is the stronger acid? (d) Why are the ions acidic? (e) Will the p$K_a$ value for [Fe(H$_2$O)$_6$]$^{2+}$ be greater or less than the value for [Fe(H$_2$O)$_6$]$^{3+}$?

2.  The alum KFe(SO$_4$)$_2$·12H$_2$O is often used in crystal growing experiments. (a) In what oxidation state is the iron centre? (b) What complex ion of iron is present in the crystal lattice of this alum? (c) What colour are crystals of KFe(SO$_4$)$_2$·12H$_2$O? (d) Chrome alum has the formua KCr(SO$_4$)$_2$·12H$_2$O; what complex ion of chromium do you think is present in crystals of chrome alum?

3.  When cobalt(II) salts dissolve in water, the species Co$^{2+}$(aq) is often shown as being present. Give the formula and structure of the cobalt-containing species that is actually present?

4.  The ion [Cr(H$_2$O)$_5$(OH)]$^{2+}$ forms a dinuclear complex in solution. (a) Write an equation that shows the formation of the dinuclear species. (b) Draw the structure of the dinuclear complex. (c) Is there any change in the oxidation state of the chromium centres upon the formation of the dichromium species?

### Case study: Aqueous vanadium chemistry

The aim of the following exercise is to provide practice in using standard reduction potentials to predict redox behaviour in aqueous solution. Answers are not provided but the relevant part of Section 16.6 in H&C will give you some useful hints, as will discussions in Section 16.9.

➤ **The hydrogen ion concentration influences the vanadium(V) solution species present**

Solutions of ammonium vanadate [NH$_4$][VO$_3$] in dilute sulfuric acid contain the [VO$_2$]$^+$ ion and are yellow. Reaction with sulfur dioxide results in a change to blue, and subsequent reduction with zinc amalgam gives an air-sensitive violet solution. Oxidation of this solution by air gives a green solution.

**Table 16.1**   Relevant standard reduction potentials for the case study of aqueous vanadium chemistry

| Half-cell reaction | $E^\circ$ / V |
|---|---|
| $Zn^{2+}(aq) + 2e^- \rightleftharpoons Zn(s)$ | $-0.76$ |
| $V^{3+}(aq) + e^- \rightleftharpoons V^{2+}(aq)$ | $-0.26$ |
| $[SO_4]^{2-}(aq) + 4H^+(aq) + 2e^- \rightleftharpoons H_2SO_3(aq) + H_2O(l)$ | $+0.17$ |
| $[VO]^{2+}(aq) + 2H^+(aq) + e^- \rightleftharpoons V^{3+}(aq) + H_2O(l)$ | $+0.34$ |
| $[VO_2]^+(aq) + 2H^+(aq) + e^- \rightleftharpoons [VO]^{2+}(aq) + H_2O(l)$ | $+0.99$ |
| $O_2(g) + 4H^+(aq) + 4e^- \rightleftharpoons 2H_2O(l)$ | $+1.23$ |

Using the data given in the last paragraph and the reduction potentials in Table 16.1, answer the following questions.

(a) Suggest identities for the coloured compounds.

(b) Is there any ambiguity in your answers?

(c) Are all of the compounds thermodynamically stable in aqueous solution?

(d) Write balanced equations for all the reactions you describe.

## COMPLEX FORMATION AND STABILITY CONSTANTS

**Problem set 16.3**

**All ligand abbreviations used in this problem set are defined in Table 16.2 of H&C**

1.   Suggest possible products (including isomers where appropriate) for the following reactions which *are* balanced on the left-hand sides:

(a) $[Cr(H_2O)_6]Cl_3 + 6KNCS \xrightarrow{\text{heat}}$

(b) $[Ni(H_2O)_6]^{2+}(aq) + 2[C_2O_4]^{2-} \rightleftharpoons$

(c) $Zn(CN)_2(s) + 2KCN(aq) \rightleftharpoons$

(d) $[Fe(H_2O)_6]^{2+}(aq) + 3bpy \rightleftharpoons$

(e) $[Fe(H_2O)_5(OH)]^{2+}(aq) + HNO_3(aq) \rightleftharpoons$

2.   Explain what is meant by the *chelate effect*. How many five-membered chelate rings are formed at the cobalt(III) centre during the formation of (a) $[Co(en)_3]^{3+}$, (b) $[Co(edta)]^-$ and (c) $[Co(dien)_2]^{3+}$ ?

3.   The complex of formula $CoBr(NH_3)_5(SO_4)$ exists as two isomers **A** and **B**. When aqueous barium chloride is added to **A**, a white precipitate is formed but no precipitate forms when the test is repeated with **B**. On the other hand, the reaction between **B** and silver nitrate solution gives a cream precipitate, while no such precipitate forms with **A**. Rationalize these observations and state in what ways isomers **A** and **B** differ.

4.   Consider the following reaction:

$$2[Co(H_2O)_6]Cl_2 + 2[NH_4]Cl + 10NH_3 + H_2O_2 \xrightarrow{\text{charcoal}} 2[Co(NH_3)_6]Cl_3 + 14H_2O$$

(a) In what oxidation state is the cobalt centre in $[Co(H_2O)_6]Cl_2$ and in $[Co(NH_3)_6]Cl_3$ ? (b) What is the role of the hydrogen peroxide? (c) If charcoal is not present, the products include $Co(NH_3)_5Cl_3$ and $Co(NH_3)_5(H_2O)Cl_3$; suggest the nature of the complex ions present in these compounds.

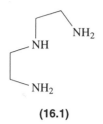

**(16.1)**

5.  In the reaction of ligand **16.1**, abbreviated to L, with aqueous nickel(II) ions, the value of log $K_1$ is 10.7, while with copper(II) ions log $K_1$ is 16.0. (a) Write down the equations to which these stability constants refer assuming that the ligand uses all of its donor atoms upon coordination to each metal centre. (b) Which of these two equilibria lies further towards the right-hand side? (c) What are the possible isomers of the product formed between nickel(II) ions and ligand **16.1** in the reaction in part (a)?

6.  For the reaction of iron(II) with cyanide ion in aqueous solution, the value of log $\beta_6$ is 24. (a) Determine the value of $\beta_6$. (b) Write down an expression for $\beta_6$ in terms of the equilibrium concentrations of $Fe^{2+}$, $[CN]^-$ and the appropriate complex ion.

7.  Stability constant data for the formation of $[MnF_6]^{4-}$ from $[Mn(H_2O)_6]^{2+}$ include the following: log $K_1 = 5.52$, log $\beta_2 = 9.02$. Determine a value for $K_2$ and comment on the relative positions of the equilibria describing the first two steps in the formation of $[MnF_6]^{4-}$.

## CRYSTAL FIELD THEORY, THE SPECTROCHEMICAL SERIES AND MAGNETISM

**Problem set 16.4**

1.  Outline the basic principles of *crystal field theory* and explain how it accounts for the fact that some, but not all, iron(II) complexes are paramagnetic.

2.  What is meant by a *strong field ligand* ? Classify the following ligands in terms of their field strength: (a) $I^-$ ; (b) $[CN]^-$ ; (c) $Cl^-$ ; (d) $H_2O$.

3.  Place the following ligands in order of *increasing* field strength: $[OH]^-$, $[CN]^-$, $Br^-$, $H_2O$, $Cl^-$ and $NH_3$.

4.  Write down the electronic configurations of the $3d$ electrons in the following complex ions: (a) low-spin $[Fe(CN)_6]^{3-}$ ; (b) high-spin $[Cr(en)_3]^{2+}$ ; (c) $[Cr(CN)_6]^{3-}$ ; (d) $[Ti(H_2O)_6]^{3+}$ ; (e) $[V(ox)_3]^{3-}$. You should also be able to represent the configurations diagrammatically as in Figure 16.23 in H&C. Why is it unnecessary to specify whether (c)–(e) are high- or low-spin ?

5.  Determine the spin-only magnetic moments of the following complexes: (a) high-spin *trans*-$[CrBr_2(en)_2]$; (b) $[VCl_4(phen)]$; (c) $[Ni(acac)_3]^-$.

6.  Is $[Sc(acac)_3]$ paramagnetic or diamagnetic? Explain your answer.

7.  The complex $[Cr(phen)_3]^{n+}$ is diamagnetic. Suggest the likely value of $n$.

8.  The *observed* magnetic moments of two complexes of the type $[M(acac)_3]$ are 2.87 and 1.74 BM. If metals M are in the first row of $d$-block, suggest possible identities for the two complexes assuming that metals M are *not* in uncommon oxidation states.

9.  The ion $[FeF_6]^{3-}$ is high-spin. (a) Draw an orbital energy diagram to show the arrangement of the $3d$ electrons in $[FeF_6]^{3-}$. (b) Calculate the spin-only

magnetic moment for $[FeF_6]^{3-}$. (c) The observed magnetic moment of $[MnF_6]^{3-}$ is 4.92 BM. Is this complex high- or low-spin? (d) Draw an energy level diagram to show the arrangement of the $3d$ electrons in $[MnF_6]^{3-}$. (e) Are the magnetic behaviours of $[FeF_6]^{3-}$ and $[MnF_6]^{3-}$ consistent with the position of $F^-$ in the spectrochemical series?

## METAL CARBONYL COMPLEXES

### The 18-electron rule

In Section 16.14 of H&C, we discussed the way in which the CO molecule acts as a ligand to low oxidation state $d$-block metals. Some of the examples we considered were $[Ni(CO)_4]$, $[Fe(CO)_5]$ and $[Cr(CO)_6]$.

A CO ligand is a 2-electron donor and in $[Ni(CO)_4]$, the total electron count at the nickel(0) centre is 18. This total is achieved by contributions from 4 CO ligands and the nickel(0) centre which has a valence electronic configuration of $4s^2 3d^8$. The nickel centre in $[Ni(CO)_4]$ *obeys the 18-electron rule*, a rule that is analogous to the octet rule that is obeyed by the lighter *p*-block elements. Instead of needing 8 electrons to fill the valence shell (e.g. for N, O or F), a *d*-block metal needs 18 electrons to fill its valence shell of one *s*, three *p* and five *d*-orbitals. Although we have not delved deeply into the theory behind it, you should recognize that the 18-electron rule *is* commonly obeyed only by *low oxidation state d-block metal centres* in organometallic compounds.

*Exercise*: Show that the each metal centre in the following species obeys the 18-electron rule: $[Cr(CO)_6]$, $[Fe(CO)_5]$, $[Mn(CO)_6]^+$, $[V(CO)_6]^-$, $[Fe(CO)_4]^{2-}$, $[Co(CO)_4]^-$.

**Problem set 16.5**

1.  What is the Dewar-Chatt-Duncanson model of bonding and to what type of compound is it applicable? Give *one* piece of experimental evidence that supports the model.

2.  Using the Kepert model, predict the shapes of the following species: (a) $[V(CO)_6]^+$; (b) $[Fe(CO)_4]^{2-}$; (c) $[Mn(CO)_5]^-$; (d) $[Ni(CO)_4]$.

3.  Nickel carbonyl is a gas at 298 K and is highly toxic. Suggest a reason for the toxicity.

4.  (a) Draw the structure of $[Fe(CO)_5]$. (b) The $^{13}C$ NMR spectrum of a solution of $[Fe(CO)_5]$ shows only one signal even at low temperature. Rationalize this observation.

5.  Carbonyl ligands can be displaced by other 2-electron ligands such as $PMe_3$. (a) Draw the structure of $PMe_3$. (b) Predict the shapes of $[Ni(CO)_3(PMe_3)]$ and $[Cr(CO)_4(PMe_3)_2]$. (c) How many isomers of $[Cr(CO)_4(PMe_3)_2]$ are there?

# 17 Carbonyl compounds

## Topics

• Physical properties and spectroscopy

• Keto-enol tautomers

• Acidity

• Syntheses and reactions

• Multi-step synthesis

## PHYSICAL PROPERTIES AND SPECTROSCOPY

**Problem set 17.1**

1.  Draw the structures of the compounds (a) propanal, (b) pentanamide, (c) hexane-2,4-dione, (d) butanedioic acid and (e) acetyl chloride.

2.  A molecular mass determination for acetic acid gives a value of 120. Rationalize this observation. [$A_r$ C = 12; O = 16; H = 1]

3.  Although methyl methanoate and acetamide have similar values of $M_r$ they have significantly different boiling points (305 and 494 K respectively). Suggest a reason for this observation.

4.  Two carbonyl compounds labelled **A** and **B** smell mousy and fruity respectively whilst a third compound **C** has an acrid odour. The three compounds are propanoyl chloride, acetamide and octyl acetate. (a) Which compound is which? (b) What causes **C** to possess an acrid smell?

5.  Figure 17.1 shows the IR spectrum of octan-3-one. (a) Which class of compound is octan-3-one? (b) Assign as many absorptions in the spectrum as you are able.

6.  (a) Draw the structure of cyclohexadiene-1,4-dione. (b) The $^{13}C$ NMR spectrum of this compound has two signals at $\delta$+136 and +187. Assign the spectrum and state the expected relative integrals of the signals.

**17.1** The IR spectrum of octan-3-one.

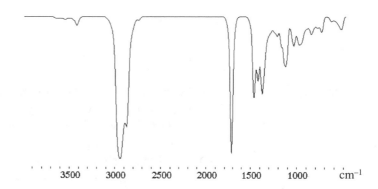

7.    Predict the multiplicities and relative integrals of the signals in the $^1$H NMR spectrum of ethyl acetate.

8.    The $^1$H NMR spectrum of a compound **A** consists of a triplet ($\delta$ 1.1, $J_{HH}$ 8 Hz), a quartet ($\delta$ 2.2, $J_{HH}$ 8 Hz) and a broad signal ($\delta$ 6.5). Compound **A** ($M_r = 73$) is a solid at 298 K. Suggest a possible identity for **A**.

## KETO-ENOL TAUTOMERS

*Tautomerism* involves the transfer of a proton between sites in a molecule with an associated rearrangement of multiple bonds. Equation 17.1 shows the equilibrium between the *keto-* and *enol*-tautomers of acetaldehyde. The equilibrium lies well to the left-hand side and we often discount the presence of the *enol*-tautomer. This is a general result: *keto*-forms usually predominate in such equilibria.

$$H-\overset{\overset{\displaystyle H}{|}}{\underset{\underset{\displaystyle H}{|}}{C}}-\overset{\overset{\displaystyle H}{}}{\underset{\underset{\displaystyle O}{\|}}{C}} \;\; \rightleftharpoons \;\; \overset{\overset{\displaystyle H}{\diagdown}\;\;\overset{\displaystyle H}{\diagup}}{\underset{\underset{\displaystyle H}{\diagup}\;\;\underset{\displaystyle OH}{\diagdown}}{C=C}} \tag{17.1}$$

**Problem set 17.2**

1.    (a) Draw the *keto*-tautomer of compound **17.1** and write an equation that shows the relationship between the two tautomers. (b) Can you write any other forms of compound **17.1**?

2.    (a) Draw the resonance forms for the deprotonated form of compound **17.1**. (b) How many $\pi$-electrons are involved in the delocalization? (c) Comparing the result in part (b) with the corresponding delocalization in acetate ion, can you formulate a general relationship between the number of electrons involved and the number of resonance forms?

3.    Using the result in question 2(c), how many resonance forms should you write for the nitrate ion?

4.    The $^{13}$C NMR spectrum of compound **17.1** is solvent dependent. The spectra in two different solvents both exhibit signals between $\delta$ 20 and 30 and also between $\delta$ 190 and 200. In solvent **A**, there is a peak at $\delta$ 58, but in solvent **B**, this is replaced by a signal at $\delta$ 100. (a) Account for these observations. (b) What additional NMR spectroscopic experiment might you do to confirm your suggestions?

**(17.1)**

## ACIDITY

**Problem set 17.3**

1.    Rationalize the *trend* in p$K_a$ values for the following acids:
$CH_3CO_2H$ (p$K_a = 4.77$), $CHCl_2CO_2H$ (p$K_a = 1.48$), $CCl_3CO_2H$ (p$K_a = 0.70$).

2.    Suggest products for the following reactions:
(a) $PhCO_2H(aq) + CsOH(aq) \rightarrow$
(b) $HO_2CCO_2H(aq) + Na_2CO_3(aq) \rightarrow$
(c) $CH_3CH_2CO_2H(aq) + [NH_4]OH(aq) \rightarrow$

3. Which of the following compounds contain α-hydrogen atoms? In the cases where you answered 'yes', identify the α-hydrogen atoms: (a) $CCl_3CO_2H$; (b) $CH_3C(O)CH_3$; (c) $HC(O)CH_2C(O)H$; (d) $Me_3CC(O)CMe_3$.

4. (a) What is the general name for a compound of the type $RC(O)CH_2C(O)R$ (where the R groups may or may not be the same)? (b) Write down the general formula and name for the conjugate base of $RC(O)CH_2C(O)R$. (c) Suggest how this conjugate base may react with iron(III) ions?

5. The $pK_a$ of $CH_3C(O)CH_2C(O)OCH_2CH_3$ is ≈ 13. If you wished to deprotonate this compound, would you choose (a) $NH_3$, (b) $NaHCO_3$ or (c) NaOH as the base? Give reasons for your choice.

6. Rationalize why a compound of type $RC(O)CH_2C(O)R$ is a stronger acid than $RC(O)CH_2R$.

## SYNTHESES AND REACTIONS

### Syntheses of carbonyl compounds

The syntheses of carbonyl compounds necessarily involve aspects of the reactivity of other carbonyl compounds. For example, the reduction of an acyl chloride gives an aldehyde, and an ester is formed by the reaction of an alcohol with a carboxylic acid.

In this set of problems, some of the reactions involve interconversions of different carbonyl compounds. We also look at ways in which spectroscopic methods can help in determining whether a particular transformation has occurred.

**Problem set 17.4**

1. How would you convert (a) butan-1-ol to butanal, and (b) butan-2-ol to butanone?

2. (a) Give the product of the reaction between butanoic acid and butan-1-ol in the presence of acid. (b) In practice, what must you do to ensure that the reaction is driven towards the products?

3. Give suitable conditions to achieve the following transformations:

4.    (a) Outline how a Grignard reagent could be used to achieve the conversion of 1-chloropentane to hexanoic acid. (b) Write a mechanism for the reaction. (c) How would the reaction be affected by a change in the starting material to 2-chloropentane? (d) Suggest an alternative method of converting 1-chloropentane to hexanoic acid.

5.    Give suitable pairs of acids and alcohols to prepare (a) PhC(O)OMe; (b) MeC(O)OCH$_2$CH$_2$Ph; (c) PhCH$_2$C(O)OMe; (d) MeC(O)OCMe$_3$.

6.    In each of the following reactions, suggest how IR spectroscopy would enable you to judge if the reaction were successful:
(a) the conversion of acetyl chloride to acetic acid;
(b) the reaction between ammonia and methyl propanoate;
(c) oxidation of ethanol to ethanoic acid;
(d) the Friedel-Crafts acylation of benzene to give ethyl phenyl ketone
(e) the acid hydrolysis of CH$_3$CH$_2$CN to CH$_3$CH$_2$CO$_2$H.

7.    Predict the appearance of the signals (*not* the chemical shifts) in the $^1$H NMR spectra of stated reactants and products in the following reactions:

(a) CH$_3$CH$_2$OH $\rightarrow$ CH$_3$CO$_2$H

(b) CH$_3$C(O)Cl $\rightarrow$ CH$_3$C(O)NH$_2$

(c) CMe$_3$CH$_2$C(O)Cl $\rightarrow$ CMe$_3$CH$_2$CHO

## Reactions of carbonyl compounds

The following problems incorporate material from Sections 17.12 and 17.13 in H&C.

**Problem set 17.5**

1.    Explain why the carbon atom of a C=O group is susceptible to attack by nucleophiles.

2.    Outline how the reaction between acetone and dichlorine proceeds in the presence of hydroxide ion. Include in your answer the products and the mechanism of the reaction.

3.    The rate equation for the acid catalysed iodination of acetone is:

$$\text{Rate of reaction} = k[\text{CH}_3\text{C(O)CH}_3][\text{H}^+]$$

(a) Explain the role of H$^+$ in this reaction. (b) Why is the reaction not dependent upon the concentration of I$_2$? (c) Sketch a graph of [I$_2$] against time.

4.    (a) What type of reagents are the ethoxide and diisopropylamide ions? Write down their formulae. (b) How does diisopropylamide ion react with acetone and how would the product of this step react with 1-bromobutane?

5.    Suggest the product(s) of the reaction between MeC(O)CH$_2$C(O)CH$_2$CH$_3$ and chloroethane in the presence of ethoxide ion. What *type* of reaction has occurred?

6.    The [OH]$^-$ ion is a poor leaving group. How can the loss of an OH group be enhanced?

7.    Outline the mechanism of the acid catalysed reaction between acetamide and water. Indicate clearly why H$^+$ must be present.

8.    (a) To what general type of compound does the reductive amination of a ketone lead? (b) Give the conditions, mechanism and product of the reductive amination of propanal.

9.    Outline the synthetic uses of the aldol reaction and provide a critical assessment of the practical disadvantages of this class of reaction.

## MULTI-STEP SYNTHESIS

In Section 14.4 of the workbook, we considered some multi-step pathways involving reactions of alkanes, alkenes and polar molecules. Now we bring together some of the reactions of aromatic and carbonyl compounds to design further multi-step syntheses. Each question has more than one possible answer and you should check your answers by reading the relevant parts of Chapters 15 and 17 of H&C. The questions also assume a knowledge of Chapters 8 and 14 of the main text.

**Problem set 17.6**

1.    Suggest a method of preparing diphenyl ketone starting from benzene as the only aromatic reagent.

2.    How would you undertake the following transformations? You may use as many steps as you think are necessary, but side products should be minimized.

(a)

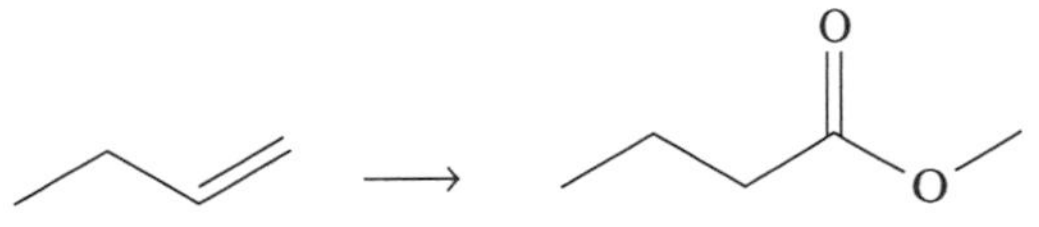

(b)

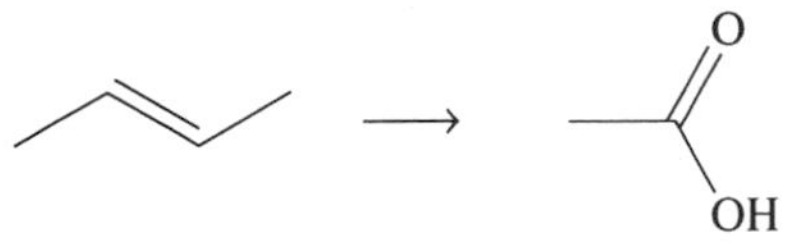

(c)

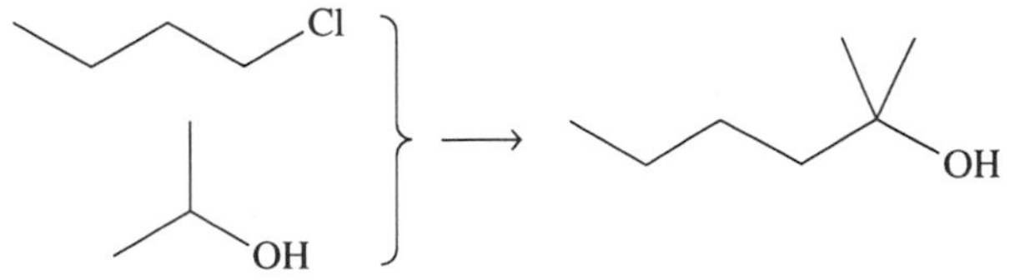

(d)

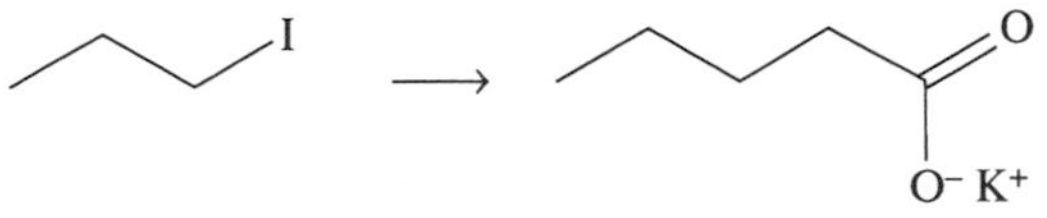

(e)

3.    How would you undertake the following transformations? You may use as many steps as you think are necessary, but point out where side products are likely to complicate the reaction pathways.

(a)

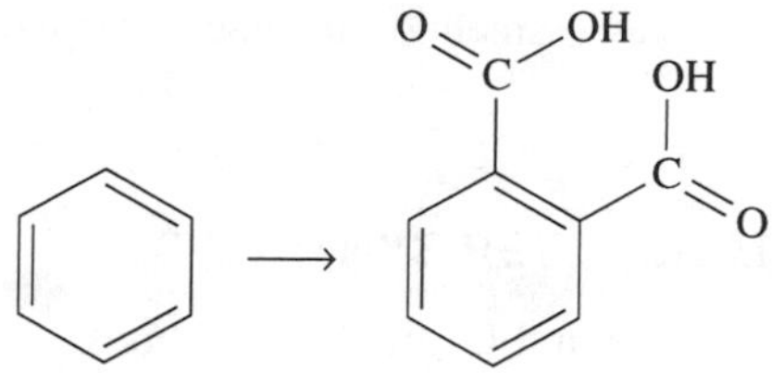

(b)

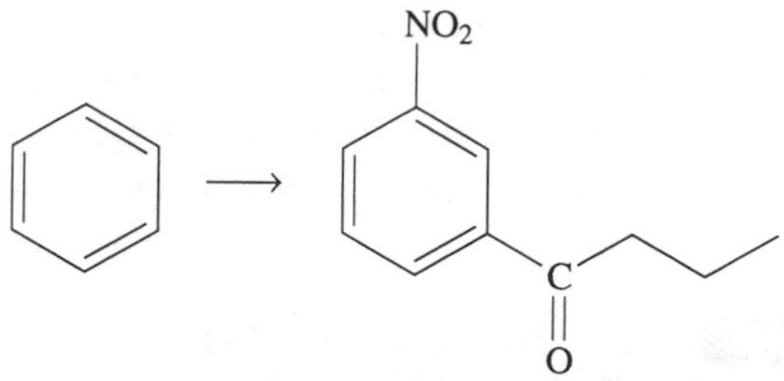

(c)

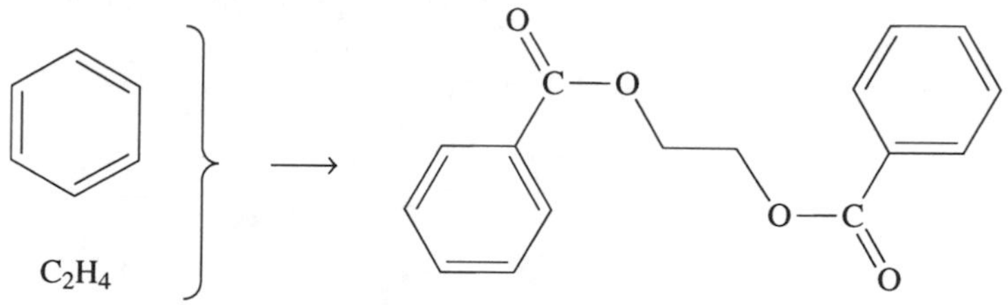

# Answers to problem sets

**Problem set 1.1**

1. Highest mass peaks of:

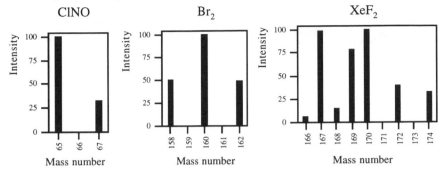

2. $^{m}/_{z}$ 38 $F_2^+$; fragmentation peak $^{m}/_{z}$ 19 $F^+$ (after F–F bond cleavage).
3. The isotope distributions of the parent ions are different. The molecules would fragment in different ways; suggest what these might be.
4. 39.13
5. 152 $(^{12}C)(^{35}Cl)_4^+$, 153 $(^{13}C)(^{35}Cl)_4^+$, 154 $(^{12}C)(^{35}Cl)_3(^{37}Cl)^+$ etc. The abundances of the ions $(^{13}C)(^{35}Cl)(^{37}Cl)_3^+$ and $(^{13}C)(^{37}Cl)_4^+$ are too low to give visible peaks. Further sets of peaks may be expected for the molecular ions $CCl_3^+$, $CCl_2^+$ and $CCl^+$ as well as $Cl^+$ and $C^+$, each showing appropriate isotopic distributions. Expected patterns are not always observed because of the differing stabilities of the ions in the mass spectrometer.

**Problem set 1.2**

1. 1.17 g
2. $1.12 \times 10^{-2}$ g (11.2 mg)
3. (a) $1 \times 10^{-3}$ mol; (b) $2.36 \times 10^{-5}$ mol
4. (a) 1.2 g; (b) 8.52 g; (c) 7.68 g; (d) 7.485 g; (e) 4.92 g; (f) 3.72 g; (g) 4.29 g

**Problem set 1.3**

1. (a) $1.25 \times 10^{-4}$ mol; (b) 3.5 mol; (c) 0.05 mol
2. (a) He $7.5275 \times 10^{22}$ atoms; (b) $H_2$ $1.204 \times 10^{23}$ atoms; (c) $S_8$ $9.635 \times 10^{18}$ atoms; (d) Na $1.807 \times 10^{23}$ atoms
3. $5.99 \times 10^{23}$ mol$^{-1}$
4. $4.666 \times 10^{-23}$ g (from $A_r/L$)

**Problem set 1.4**

1. 61.9 dm$^3$
2. $\Delta V = -10.05$ dm$^3$
3. $V(400\text{ K}) = 67.8$ dm$^3$; $\Delta V = 17.8$ dm$^3$
4. 0.40 dm$^3$
5. New $P = 0.89 \times 10^5$ Pa; $\Delta P = -(0.11 \times 10^5)$ Pa

6.  (a) 2.65 mol [$V$(273 K) = 60.06 dm³]; (b) no; (c) $P_{H_2}$ = 0.15 bar, $P_{He}$= 0.64 bar

7.  $V$(273 K, 10⁵ Pa) = 1.39 dm³; mass = 2.69 g

8.  $1.76 \times 10^{-3}$ mol

9.  (a) Increase $T$ from 290 K to 441 K at constant $P$; (b) decrease $P$ from 1 bar to 0.657 bar at constant $T$.

---

### Problem set 1.5

1.  Solid $NH_4Cl$ first forms nearer to the right-hand (HCl) end of the tube; $M_r(NH_3)$ = 17 and $M_r(HCl)$ = 36.5; by Graham's Law, rate of diffusion of $NH_3$ is $\approx$1.5 times that of HCl.

2.  $\dfrac{2.25}{1.4} = \sqrt{\dfrac{M_r(\mathbf{X})}{17}}$; $M_r(\mathbf{X})$ = 44 (gas could be $CO_2$ or $C_3H_8$)

---

### Problem set 1.6

1.  0.01 mol
2.  $1.25 \times 10^{-2}$ mol
3.  0.08 mol
4.  0.033 mol
5.  0.2 M; in 250 cm³ there are 0.05 mol K⁺ ions
6.  2.00 g (0.025 mol)
7.  1.46 g (0.025 mol)
8.  0.04 M
9.  0.5 M

---

### Problem set 1.7

1.  $CaH_2(s) + 2H_2O(l) \rightarrow Ca(OH)_2(aq) + 2H_2(g)$

    $2Na(s) + 2H_2O(l) \rightarrow 2NaOH(aq) + H_2(g)$

    $2C_6H_{14}(l) + 19O_2(g) \rightarrow 12CO_2(g) + 14H_2O(l)$

    $SnO_2(s) + 2KOH(aq) + 2H_2O(l) \rightarrow K_2[Sn(OH)_6](aq)$

    $Pb(NO_3)_2(aq) + 2NaCl(aq) \rightarrow PbCl_2(s) + 2NaNO_3(aq)$

    $MnO_2(s) + 4HCl(aq) \rightarrow MnCl_2(aq) + Cl_2(g) + 2H_2O(l)$

2.  (a) 0.05 mol Na = 1.15 g; (b) 0.01 mol $PbCl_2$ = 2.78 g; (c) 4.8 mol $CO_2$ = 118.9 dm³; (d) 0.1 mol $H_2$ = 2.45 dm³ at 295 K; (e) 0.15 mol $MnO_2$ = 13.05 g

3.  $V$(273 K) = 72.6 dm³; $V$(300 K) = 79.8 dm³

---

### Problem set 1.8

1.  (a) H +1, S –2; (b) C +4, O –2; (c) S +4, O –2, Cl –1; (d) Ca +2, F –1; (e) Mn +4, O –2; (f) Ge +4, Cl –1; (g) S +6, F –1; (h) P +5, O –2, Cl –1; (i) K +1, Mn +7, O –2; (j) C +4, F –1, Cl –1; (k) H +1, O –1 (homonuclear O–O bond does not contribute to oxidation state calculation); (l) S 0; (m) C 0; (n) Fe +3, O –2

➤

**[BH₄]⁻ is a *hydride*: see H&C Section 11.3**

2.  (a) N +3, O –2; (b) Cr +6, O –2; (c) Sn + 4, Cl –1; (d) H +1, O –2; (e) B +3, H –1; (f) Cl +7, O –2; (g) Cl +5, O –2

---

**Problem set 1.9**   1–2.  See table below.

| Reaction | Species oxidized (change in oxidation state) | Species reduced (change in oxidation state) | Oxidizing agent | Reducing agent |
|---|---|---|---|---|
| (a) | K (0 to +1) | H (+1 to 0) | $H_2O$ | K |
| (b) | Mg (0 to +2) | H (+1 to 0) | HCl | Mg |
| (c) | Al (0 to +3) | Fe (+3 to 0) | $Fe_2O_3$ | Al |
| (d) | N (0 to +4) | O (0 to –2) | $O_2$ | $N_2$ |
| (e) | H (0 to +1) | Cl (0 to –1) | $Cl_2$ | $H_2$ |
| (f) | Cu (0 to +2) | Cr (+6 to +3) | $[Cr_2O_7]^{2-}$ | Cu |
| (g) | P (0 to +5) | O (0 to –2) | $O_2$ | $P_4$ |
| (h) | H (0 to +1) | Si (+4 to 0) | $SiCl_4$ | $H_2$ |
| (i) | O (–1 to 0) | Cl (0 to –1) | $Cl_2$ | $H_2O_2$ |
| (j) | I in $I^-$ (–1 to 0) | I in $[IO_3]^-$ (+5 to 0) | $[IO_3]^-$ | $I^-$ |

**Problem set 1.10**

1.  (a) –1113 kJ per mole of reaction;  (b) –198 kJ per mole of reaction; (c) –1344 kJ per mole of reaction; (d) –115.5 kJ per mole of reaction; (e) –852 kJ per mole of reaction; (f) –184 kJ per mole of reaction; (g) –1367 kJ per mole of reaction (this is also $\Delta_c H^\circ$(298 K)); (h) –162.5 kJ per mole of reaction (notice this was *gaseous* $Br_2$).

2.  $\Delta_f H^\circ$(298 K) = –248.5 kJ per mole of acetone

3.  (a) $\Delta H(1)$ = –175.5 kJ per mole of $C_2H_2$; $\Delta H(2)$ = –136.5 kJ per mole of $C_2H_4$; $\Delta H(3)$ = –312 kJ per mole of $C_2H_2$; (b) By Hess's Law:
$$\Delta H(1) + \Delta H(2) = \Delta H(3)$$
and substitution of the values into the equation confirms the law.

**Problem set 1.11**

1.  Exothermic reaction in the forward direction, and raising the temperature favours the back-reaction; heat energy is taken in from the surroundings lowering the temperature again. Increasing the pressure favours the forward reaction; 4 moles → 2 moles and fewer moles of gases reduces the pressure.

2.  Endothermic reaction in the forward direction and raising the temperature favours the forward-reaction; heat energy is taken in from the surroundings lowering the temperature again. Increasing the pressure favours the forward reaction; 3 moles → 2 moles and fewer moles of gases reduces the pressure.

3.  Adding $[OH]^-$ pushes the equilibrium to the right-hand side so it is used up.

4.  (a) Adding propanoic acid pushes the equilibrium to the right-hand side so it is consumed; (b) removing benzyl propanoate pushes the equilibrium to the right-hand side so that more is formed.

**Problem set 1.12**

1.
$$CH_3CO_2H(aq) + H_2O(l) \rightleftharpoons [CH_3CO_2]^-(aq) + [H_3O]^+(aq)$$

Initial moles:    0.4    excess    0    0

Equilm moles: $0.4 - (2.6 \times 10^{-3})$    excess    $2.6 \times 10^{-3}$    $2.6 \times 10^{-3}$

$K_c = 1.7 \times 10^{-5}$ mol dm$^{-3}$ *without* the assumption $[CH_3CO_2^-]_{equilm} \approx [CH_3CO_2^-]_{initial}$

2.
$$I_2(aq) \quad + \quad I^-(aq) \quad \rightleftharpoons \quad [I_3]^-(aq)$$

Initial moles:    0.01    0.5    0

Equilm moles: $0.01 - (9.8 \times 10^{-3})$    $0.5 - (9.8 \times 10^{-3})$    $9.8 \times 10^{-3}$

$K_c = 100$ mol$^{-1}$ dm$^3$

3.
$$CH_3CO_2H(l) + CH_3CH_2OH(l) \rightleftharpoons CH_3CO_2C_2H_5(l) + H_2O(l)$$

Initial moles:    1    0.5    0    0

Equilm moles:   $(1 - 0.425)$    $(0.5 - 0.425)$    0.425    0.425

$K_c = 4.19$ (dimensionless)

4.
$$H_3PO_3(aq) + H_2O(l) \rightleftharpoons [H_3O]^+(aq) + [H_2PO_3]^-(aq)$$

Initial moles:    0.1    excess    0    0

Equilm moles:   $(0.1 - x)$    excess    $x$    $x$

$[H_2PO_3]^- = 2.7 \times 10^{-2}$ mol dm$^{-3}$ *without* assumption $[H_2PO_3^-]_{equilm} \approx [H_2PO_3^-]_{initial}$

**Problem set 1.13**

1.
$$N_2(g) \quad + \quad 3H_2(g) \quad \rightleftharpoons \quad 2NH_3(g)$$

Initial moles:    0.5    2    0

Equilm moles:   $(0.5 - x)$    $(2 - 3x)$    $0.81 = 2x$

$K_p = 40.7$ bar$^{-2}$

2.
$$2SO_2(g) \quad + \quad O_2(g) \quad \rightleftharpoons \quad 2SO_3(g)$$

Initial moles:    2    0.5    0

Equilm moles:   $(2 - 2x)$    $(0.5 - x)$    $0.24 = 2x$

$K_p = 0.13$ bar$^{-1}$

3.
$$H_2(g) \quad + \quad I_2(g) \quad \rightleftharpoons \quad 2HI(g)$$

Initial moles:    1    1    0

Equilm moles:   $(1 - x)$    $(1 - x)$    $2x$

$x = 0.86$; moles of HI at equilibrium $= 1.72$.

4.
$$H_2(g) \quad + \quad CO_2(g) \rightleftharpoons H_2O(g) \quad + \quad CO_2(g)$$

Initial moles:    0.8    0.6    0    0

Equilm moles:   $(0.8 - x)$    $(0.6 - x)$    $x$    $x$

$x = 0.24$; moles of CO$_2$ remaining at equilibrium $= 0.36$.

**Problem set 2.1**

1.  (a) $p$ ; (b) $f$ ; (c) $d$ ; (d) $s$.
2.  7 ($l = 3$; $m_l = +3, +2, +1, 0, -1, -2, -3$)
3.  $n = 4, l = 0, m_l = 0$ corresponds to a $4s$ AO; $n = 4, l = 1, m_l = +1, 0, -1$ corresponds to three $4p$ AOs; $n = 4, l = 2, m_l = +2, +1, 0, -1, -2$ corresponds to five $4d$ AOs; $n = 4, l = 3, m_l = +3, +2, +1, 0, -1, -2, -3$ corresponds to seven $4f$ AOs. Total number of AOs with $n = 4$ is 16.
4.  $2p$ AOs labelled by the three sets of quantum numbers [$n = 2, l = 1, m_l = +1$]; [$n = 2, l = 1, m_l = 0$]; [$n = 2, l = 1, m_l = -1$].
5.  (a) $n = 1$, only the $1s$ AO; (b) $n = 2$, two subshells, $2s$ and $2p$; (c) $n = 3$, three

sub-shells, 3*s*, 3*p*, 3*d*; (d) *n* = 4, four subshells, 4*s*, 4*p*, 4*d*, 4*f*.

6.   (a), (c) and (e) cannot exist.

7.   $m_s$

8.   $[n = 4, l = 1, m_l = +1, m_s = +\frac{1}{2}]$; $[n = 4, l = 1, m_l = +1, m_s = -\frac{1}{2}]$; $[n = 4, l = 1, m_l = 0, m_s = +\frac{1}{2}]$; $[n = 4, l = 1, m_l = 0, m_s = -\frac{1}{2}]$; $[n = 4, l = 1, m_l = -1, m_s = +\frac{1}{2}]$; $[n = 4, l = 1, m_l = -1, m_s = -\frac{1}{2}]$.

9.   3*s* orbital; one of the two electrons has the set of quantum numbers $[n = 3, l = 0, m_l = 0, m_s = +\frac{1}{2}]$, the other has the set $[n = 3, l = 0, m_l = 0, m_s = -\frac{1}{2}]$.

---

**Problem set 2.2**

1-2.   See H&C: Section 2.11.

3.   (a) None; (b) one.

4.   Increasing size: 1*s* < 2*s* < 3*s* < 4*s* < 5*s*; 5*s* is the most diffuse.

5.   (a) Most diffuse: 5*p*; (b) least diffuse: 2*p*.

6.   Increasing energy: 1*s* < 2*s* < 2*p* < 3*s* < 3*p*.

---

**Problem set 2.3**

1-2.   See H&C: Section 2.13.

3.   See Figure 2.3.1.

**2.3.1** Radial distribution curves for 3d, 4p and 4s atomic orbitals.

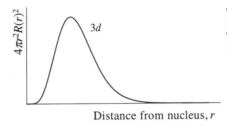

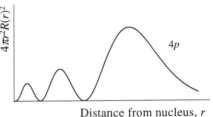

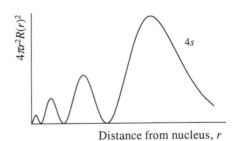

4.   Each curve indicates the probability of finding an electrons at distance *r* from the nucleus (at *r* = 0). Points where the function $4\pi r^2 R(r)^2$ are at a maximum correspond to maximum probability — the position where the electron is most likely to spend its time. For the 3*s* AO, the function $4\pi r^2 R(r)^2$ has three maxima, corresponding to three regions of high probability of finding the electron; by comparing the three maxima, you should see that the electron is *most* likely to be further away from the nucleus than close to it.

1. Hydrogen-like species must possess *one electron*. The atomic numbers of H, He and Li are 1, 2 and 3. Only (d) $He^+$ and (f) $Li^{2+}$ are hydrogen-like.

2. Orbital energy decreases (more negative) as the nuclear charge increases.

3. Orbital size decreases (it *contracts*) as the nuclear charge increases.

4. See Figure 2.4.1.

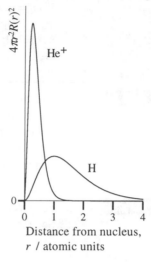

**2.4.1** Radial distribution functions for the $1s$ atomic orbitals in the H atom and $He^+$ ion.

---

1. Your answer should look similar to the $1s$ curve in Figure 2.2, page 19.

2. See H&C: Section 2.15.

3. $2s$ is more penetrating than $2p$.

4. $2s$ AO; the $2s$ AO penetrates closer to the nucleus than the $2p$ orbital and an electron is the $2s$ orbital feels a greater effective nuclear charge than an electron in the $2p$ AO; the energy of the electron in the $2s$ AO is lowered with respect to it being in a $2p$ AO.

5. See Figure 2.5.1. In the hydrogen atom, the $2s$ and $2p$ levels are degenerate (have the same energy).

6. See H&C: Section 2.15.

7. Absolute nuclear charge is constant for a given nucleus or ion and is governed by the number of protons. The effective nuclear charge is a variable; electrons in different orbitals are attracted to the nucleus by different amounts; see H&C: Section 2.15.

$E_{AO}$   —  —  —  $2p$

— $2s$

— $1s$

**2.5.1** Approximate orbital energies in lithium.

---

1. $\Delta n = 0, 1, 2, 3...$; $\Delta l = \pm 1$

2. In atomic H, $2s$ and $2p$ have the same energy (degenerate); there is no change in energy associated with the $2p \rightarrow 2s$ transition.

3. (a) Lyman; (b) Balmer; (c) Balmer; (d) Lyman.

4. See Figure 2.6.1; each line in the spectrum corresponds to a different transition; the transition for which the energy change is the largest $(2 \rightarrow 1)$ corresponds to the highest frequency line in the spectrum.

5. $E = h\nu$; $E = 2.001 \times 10^{-18}$ J. [How do you convert this to kJ mol$^{-1}$ ?]

6. See Figure 2.6.2.

7. Emission: (high energy level) $\rightarrow$ (low energy level); absorption: (high energy level) $\leftarrow$ (low energy level); see H&C: Box 2.7.

8.    In ground state, the electron in H atom is in the $n = 1$ level; when promoted to $n = \infty$, it can escape, i.e. the atom is ionized. The ionization energy is defined for the gas phase process:

$$H(g) \rightarrow H^+(g) + e^-$$

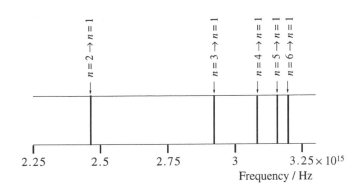

**2.6.1** Part of the spectrum of atomic hydrogen showing lines in the Lyman series.

**2.6.2** Selected transitions in the Balmer and Lyman series.

---

**Problem set 2.7**

1.    See Figure 2.7.1; gradient $= R = 3.90 \times 10^{15}$ Hz.

2.    (a) $n = 2.983 \times 10^{14}$ Hz; $E = 1.977 \times 10^{-19}$ J or $1.190 \times 10^2$ kJ mol$^{-1}$; (b) $2.741 \times 10^{14}$ Hz; $E = 1.816 \times 10^{-19}$ J or $1.094 \times 10^2$ kJ mol$^{-1}$; (c) $2.339 \times 10^{14}$ Hz; $E = 1.550 \times 10^{-19}$ J or 93.34 kJ mol$^{-1}$; (d) $1.599 \times 10^{14}$ Hz; $E = 1.059 \times 10^{-19}$ J or 63.77 kJ mol$^{-1}$; infrared part of spectrum.

3.    (a) No; (b) no; Lyman series gives rise to emission lines in the ultraviolet, and Balmer in the visible parts of the spectrum.

**2.7.1** Using the Lyman series to find the Rydberg constant.

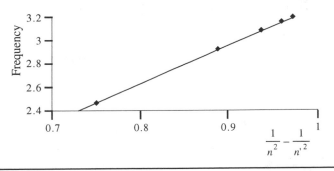

---

**Problem set 2.8**

1.    (a) $1s^2 2s^2 2p^3$; (b) $1s^2 2s^2$; (c) $1s^2 2s^2 2p^6 3s^2 3p^5$; (d) $1s^2 2s^2 2p^6 3s^2 3p^1$; (e) $1s^2 2s^2 2p^6 3s^2 3p^6 4s^1$.

2.    (a) $1s^2 2s^2 2p^6$; (b) $1s^2 2s^2 2p^6$; (c) $1s^2 2s^2 2p^6 3s^2 3p^6$; (d) $1s^2$. Each has a noble gas configuration.

3.    See Figure 2.8.1.

4.    Ground state configuration is $1s^2 2s^2 2p^4$ ; the $1s$ electrons are core and the $2s$ and $2p$ electrons are in the valence shell.

5.    The ground state configuration is $1s^2 2s^2 2p^5$, and there are seven valence electrons.

6.    One; the valence electronic configuration is $ns^2np^1$; more generally, the ground state electron configuration is $[X]ns^2np^1$ where X is the noble gas preceeding the group 1 element in the periodic table.

7.    Seven valence electrons; a valence electronic configuration of $ns^2np^5$.

8.    Four valence electrons; a valence electronic configuration of $ns^2np^2$.

**2.8.1** Ground state electronic configurations.

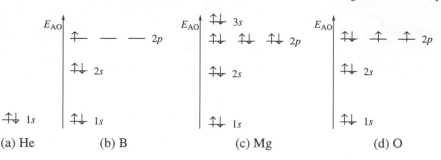

(a) He          (b) B          (c) Mg          (d) O

---

**Problem set 2.9**

1.    See H&C: Section 2.20.

2.    (a) Loses one electron; (b) gains one electron; (c) loses three electrons; (d) gains two electrons; (e) loses two electrons.

3.    (a) $Na^+$; (b) $O^{2-}$; (c) $Cl^-$; (d) $Mg^{2+}$; (e) $N^{3-}$.

4.    (a) C will *share* four electrons; (b) N will *share* three electrons (in only relatively few compounds does it *gain* three electrons); (c) F will *gain* one electron; (d) Na will *lose* one electron; (e) Cl will *gain* one electron.

5.    $SCl_2$

6.    ClF

7.    (a) $Na_3N$; (b) NaF; (c) $MgCl_2$; (d) $NF_3$; (e) $OF_2$.

---

**Problem set 2.10**

1.    Fully occupied valence shell orbitals (a 'noble gas configuration').

2.    Group 17

3.    Group 14

4.    Group (a) 2; (b) 16; (c) 18; (d) 1; (e) 17

5.    $ns^2np^4$

6.    $ns^1$

7.    $ns^2$

8.    $ns^2np^1$

---

**Problem set 2.11**

1.    Helium cannot be solidified; see H&C, page 89, Table 2.6.

2.    Hydrogen and elements in groups 15, 16, 17 and 18, although P and S have longer ranges than N and O in groups 15 and 16.

3.    Expanding the temperature axis (or inspection of the tabulated data) shows that the noble gases have extremely short liquid ranges, much shorter than other elements with the exception of hydrogen.

4.    Largest liquid ranges occur for elements in groups 1, 2, 13 and 14.

5.    Graphite and diamond (carbon, $Z = 6$) possess giant molecular lattices; nitrogen ($Z = 7$) is composed of diatomic molecules. Covalent bonds are broken when C melts, but only weak intermolecular interactions are

overcome when solid $N_2$ melts.

> **More about the structures of elements in Chapter 7 of H&C**

6.  Helium ($Z = 2$), neon ($Z = 10$) and argon ($Z = 18$) are monatomic gases with weak van der Waals forces between the atoms; lithium ($Z = 3$), sodium ($Z = 11$) and potassium ($Z = 19$) are metallic elements and metal–metal bonds are broken when the solid elements melt.

---

**Problem set 3.1**

1.  (a) First row of the p-block (boron to fluorine); (b) $r_v$ is half the distance between two non-bonded atoms; $r_{cov}$ is half the distance between two bonded (single bond) atoms; (c) as $Z$ increases from 5 to 9, the effective nuclear charge increases and the atom becomes smaller.

2.  $r_v$ is half the distance between two non-bonded S atoms; $r_{cov}$ is half of a single S–S bond distance. Both can be estimated from the solid state structure of an allotrope of sulfur, e.g. $S_8$ (see Table 3.2 in H&C).

3.  Estimated bond distance, $d = 2 \times r_{cov}$ : $d(F_2) = 142$, $d(Cl_2) = 198$, $d(Br_2) = 228$, $d(I_2) = 266$ pm. Bond distance increases as group 17 is descended; the further down the group, the more electrons and quantum shells the atom possesses and atomic size increases.

---

**Problem set 3.2**

1.  $D$ = bond dissociation enthalpy of a particular bond; $\overline{D}$ = average bond dissociation enthalpy (for a series of similar bonds).

2.  For $X_2$: $D(X - X) = 2 \times \Delta_a H^\circ (X)$
    The shape of an $X_2$ molecule is defined; there can only be one X–X bond. The shape of an $X_n$ molecule ($n \geq 3$), is not apparent from the formula. e.g. $P_4$ is tetrahedral (Figure 3.2, page 26) and there are six P–P bonds. For an $X_4$ ring, there are four X–X bonds. It is **not** possible to write a general equation relating $D(X–X)$ and $\Delta_a H^\circ$ for $X_n$.

3.  (a) 436, (b) 242, (c) 946, (d) 277, (e) 210 kJ mol$^{-1}$.

4.  1662 kJ per mole of $S_6$; $S_6(g) \rightarrow 6S(g)$.

5.

In the series $C_2H_2$, $C_2H_4$, $C_2H_6$, the carbon-carbon bond changes from a triple to double to single bond; $D(C\equiv C) > D(C=C) > D(C–C)$. An O=O double bond is present in $O_2$, but $H_2O_2$ possesses an O–O single bond. A $N\equiv N$ triple bond (in $N_2$) is stronger than an N–N single bond (in $N_2H_4$). Both $S_2H_2$ and $S_8$ possess S–S single bonds and the bond enthalpies are similar.

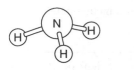

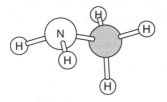

**3.3.1** The structures of $NH_3$ and $CH_3NH_2$.

1.  (a) Suitable compounds to use are ethyne ($C_2H_2$) and methane ($CH_4$). Data needed: $\Delta_aH°(C) = 717$; $\Delta_aH°(C) = 218$; $\Delta_fH°(C_2H_2,$ g$) = +228$; $\Delta_fH°(CH_4,$ g$) = -74$ kJ mol⁻¹. Determine $\overline{D}(C–H)$ using data for $CH_4$: 415.75 kJ mol⁻¹. Transfer to ethyne and estimate $D(C≡C) = 810.5$ kJ mol⁻¹.

2.  Data needed: $\Delta_aH°(P) = 315$; $\Delta_aH°(F) = 79$; $\Delta_fH°(PF_5,$ g$) = -1594$; $\Delta_fH°(PF_3,$ g$) = -958$ kJ mol⁻¹. (a) $\overline{D}(P–F)$ from $PF_5 = 460.8$ kJ mol⁻¹; (b) $\overline{D}(P–F)$ from $PF_3 = 503.3$ kJ mol⁻¹. In $PF_3$, phosphorus is in oxidation state +3, but is in oxidation state +5 in $PF_5$. $\overline{D}(P–F)$ is dependent on oxidation state.

3.  A suitable compound is $PH_3$: $\Delta_aH°(P) = 315$; $\Delta_aH°(H) = 218$; $\Delta_fH°(PH_3,$ g$) = +5$ kJ mol⁻¹; $\overline{D}(P–H) = 321.3$ kJ mol⁻¹. Using this value with data for $P_2H_4$ ($\Delta_fH°(P_2H_4,$ g$) = +21$ kJ mol⁻¹) gives $D(P–P) = 195.8$ kJ mol⁻¹. This compares with 210 kJ mol⁻¹ from $P_4$, giving an indication of the limitations of the method of bond enthalpy transferability.

4.  (a) No (C=C and C≡C); (b) yes (all $S_n$ rings with S–S single bonds, see Table 3.2 in H&C); (c) no (sulfur oxidation state is not constant; see question 2 above); (d) yes (both N–H single bonds in similar environments, Figure 3.3.1).

1.  (a) (b) (c) (d) (e) (f) (g) (h) (i)

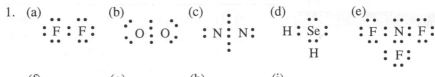

2.  $CO_2$: no lone pairs on C; $SO_2$: one lone pair on S.

3.  (a) sextet  (b) octet  (c) octet

4.  octet    octet    (10 electrons in valence shell)    (12 electrons in valence shell)

**Problem set 3.5**

1.  H **:** H

2.  H⁻ is H **:** and so [H₂]⁻ can be represented as H **:** H  •

3.  Combination of two resonance structures:   H−H ⟷ H⁺ H⁻

4.  See Figure 3.5.1.

5.  (a) Bond order = 1; (b) diamagnetic.

6.  (a) See Figures 3.5.2 and 3.5.3; (b) bond order in each = $^1/_2$; (c) ions could exist but H–H bond would be weak; (d) both ions would be paramagnetic.

7.  $\sigma$-Bond in each.

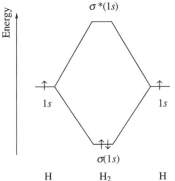

**3.5.1** MO diagram for the formation of H₂.

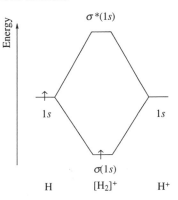

**3.5.2** MO diagram for the formation of [H₂]⁺.

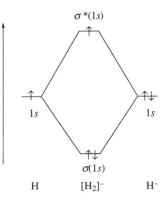

**3.5.3** MO diagram for the formation of [H₂]⁻.

**Problem set 3.6**

1.  Li **:** Li       The Lewis structure shows that the electrons are paired (diamagnetic) and there is an Li–Li single bond.

2.  Li—Li ⟷ Li⁺Li⁻       $\Psi_{molecule} = \Psi_{covalent} + [c \times \Psi_{ionic}]$
    where $c$ is a coefficient showing the contribution made by the ionic resonance structure.

3.  See Figures 3.6.1 to 3.6.5. Ground state electronic configurations: (a) Li₂ $\sigma(2s)^2$; (b) [Li₂]⁺ $\sigma(2s)^1$; (c) He₂ $\sigma(1s)^2\sigma^*(1s)^2$; (d) Be₂ $\sigma(2s)^2\sigma^*(2s)^2$; (e) [He₂]²⁺ $\sigma(1s)^2$; bond orders: (a) 1; (b) $^1/_2$; (c) 0; (d) 0; (e) 1; Li₂, [Li₂]⁺ and [He₂]²⁺ should be viable, based on the bond order being > 0.

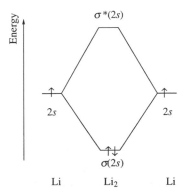

**3.6.1** MO diagram for the formation of Li₂ (valence electrons only).

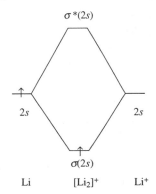

**3.6.2** MO diagram for the formation of [Li₂]⁺ (valence electrons only).

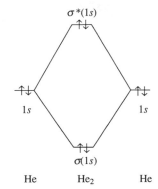

**3.6.3** MO diagram for the formation of He₂.

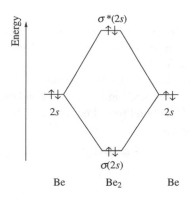

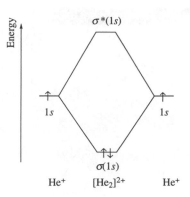

**3.6.4** MO diagram for the formation of $Be_2$ (valence electrons only).

**3.6.5** MO diagram for the formation of $[He_2]^{2+}$. Compare with Figure 3.5.1; $He^+$ has one electron like H.

4.   Bond order $[Li_2]^+ < Li_2$, and so the bond dissociation enthalpy should be less than 110 kJ mol$^{-1}$.

5.   No; not all single bonds have the same bond dissociation enthalpy. Even though Li and F are in the same row of the periodic table, factors such as internuclear repulsion (in $Li_2$) and repulsions between the lone pairs of electrons (in $F_2$) contribute to the value of $D(X-X)$.
     [$D(F-F) = 159$ kJ mol$^{-1}$.]

---

**Problem set 3.7**

1.   From Figure 3.4 (page 30): all but (c) and (d) are symmetry-allowed.
2.   (a), (b), (e) are $\sigma$-interactions; (f), (h) and (i) are $\sigma^*$-interactions; (g) is a $\pi^*$ interaction; there is no example of a $\pi$ interaction.
3.   A $\sigma$-orbital is symmetrical about the internuclear axis; a $\sigma^*$-orbital is symmetrical about the internuclear axis, but also has a nodal plane between the nuclei.
4.   See Figure 3.18 in H&C.

---

**Problem set 3.8**

1.   See H&C, Section 3.12.
2.   See Figure 3.8.1.
3.   See Figure 3.8.2.
4.   Using Figure 3.8.1: (a) bond order = 1; (b) diamagnetic; using Figure 3.8.2: (a) bond order = 1; (b) paramagnetic; $B_2$ is experimentally found to be paramagnetic.
5.   No; see H&C, Section 3.19.
6.   See H&C, Section 3.18.
7.   (a) $O_2$, $[O_2]^-$ and $[O_2]^{2-}$. Best approached using an MO treatment, to show addition of electrons to the $\pi^*$ level and bond order changing from 2 to 1.5 to 1. (b) $Li_2$, $Na_2$ and $K_2$. Each method will show an X–X single bond. The decrease in $D(X-X)$ can be interpreted in terms of the more diffuse nature (less overlap) of the 4s AOs of K compared to the 3s AOs of Na and the 2s

AOs of Li. (c) $F_2$ to $I_2$. Each method will show an X–X single bond. Consideration of the AOs involved and the proximity of the lone pairs of electrons will provide a rationalization of the trend observed.

**3.8.1** An approximate MO diagram for the formation of a diatomic $X_2$ assuming that there is no orbital mixing (no $\sigma$-$\pi$ crossover). This leads to the filling of the $\sigma(2p_z)$ MO before the $\pi(2p_x)$ and $\pi(2p_y)$ MOs. [Atom X has $2s$ and $2p$ valence orbitals.]

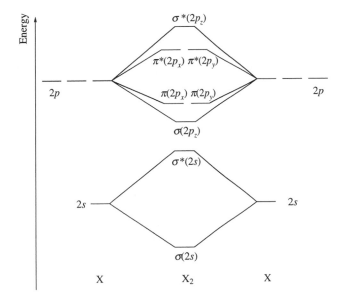

**3.8.2** An approximate MO diagram for the formation of a diatomic $X_2$ assuming that there is orbital mixing (a $\sigma$-$\pi$ crossover occurs). This leads to the filling of the $\pi(2p_x)$ and $\pi(2p_y)$ MOs before the $\sigma(2p_z)$ MO. [Atom X has $2s$ and $2p$ valence orbitals.]

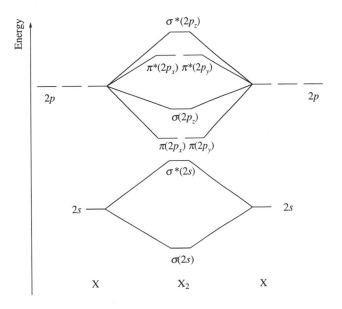

**Problem set 4.1**

1.  F—F $\longleftrightarrow$ F$^+$ F$^-$ $\longleftrightarrow$ F$^-$ F$^+$    The covalent form predominates.
2.  (c) is correct.
3.  (a) H—F $\longleftrightarrow$ H$^+$ F$^-$ $\longleftrightarrow$ H$^-$ F$^+$    The structure with F$^+$ can be ignored.
    (b) Na—H $\longleftrightarrow$ Na$^+$ H$^-$ $\longleftrightarrow$ Na$^-$ H$^+$    All should contribute to some extent.
    (c) Cl—F $\longleftrightarrow$ Cl$^+$ F$^-$ $\longleftrightarrow$ Cl$^-$ F$^+$    The covalent form is dominant; Cl$^+$ F$^-$ will contribute to some extent.

---

**Problem set 4.2**

1.  [He]$2s^22p^5$ is the ground configuration of both O$^-$ and F; [OH]$^-$ has the same number of electrons (valence and core) as HF (they are isoelectronic, see page 37 of the workbook).

2.  H $\overset{\displaystyle\cdot\cdot}{\underset{\displaystyle\cdot\cdot}{:\text{O}:}}$

3.  H$-$O$^-$ $\longleftrightarrow$ H$^+$ O$^{2-}$ $\longleftrightarrow$ H$^-$ $\overset{\displaystyle\cdot\cdot}{\underset{\displaystyle\cdot\cdot}{\text{O}:}}$    i.e. covalent and ionic forms.

    The covalent form should predominate; the right hand resonance structure can be ignored because the O atom has a sextet of valence electrons (the electrons have been included to emphasize this fact).

4.  (a) Bond order = 1; (b) diamagnetic.

5.  See Figure 4.1.1. The relative energies of the hydrogen and the oxygen atomic orbitals are not easy to assess. You know that F and O$^-$ are related, and so the orbital energies for O$^-$ may resemble those of F, *although* the negative charge of the O$^-$ ion will result in the orbital energies being raised (less negative); this is the opposite effect to increasing the effective nuclear charge (see H&C, Section 2.15). Your MO diagram may not look exactly like that in Figure 4.1.1, but provided you have obtained the correct ordering and types of MOs, then you can consider it to be a reasonable approximation.

6.  (a) Bond order = 1; (b) diamagnetic; (c) from Figure 4.1.1, the bonding MO has similar amounts of H and O character — your answer will depend on how you assessed the relative energies of the H $1s$ and O $2p$ AOs, e.g. if you drew the O $2p$ at *lower* energy than the H $1s$, then the $\sigma$-MO will have *more* O than H character.

**4.1.1** Approximate MO diagram for the formation of [OH]$^-$; the break on the energy axis indicates that the energy of the $2s$ is lower than is shown.

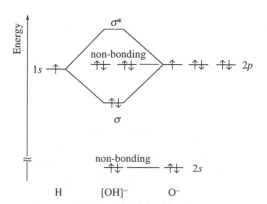

7. Similarities between MO diagrams (and conclusions drawn) for HF (H&C Figure 4.8) and $[OH]^-$: pattern of MOs, bond order, diamagnetism. Differences: relative energies of the H and $O^-$, or H and F, AOs are not the same, resulting in a difference in the characters of the $\sigma$-bonding MOs.

---

**Problem set 4.3**

1. See Figure 4.3.1

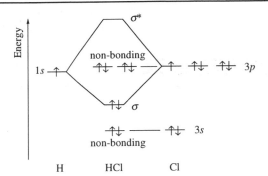

**4.3.1** MO diagram for the formation of HCl using an LCAO approach.

2. (a) Bond order = 1; (b) diamagnetic; (c) and (d) ≈ equal contributions from H $1s$ and Cl $3p_z$ (if the H and Cl nuclei to lie on the $z$ axis). Representations of the $\sigma$ and $\sigma^*$ MOs are:

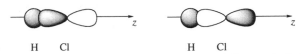

3. You should include in your answer the following: both $1s$–$3s$ and $1s$–$3p_z$ orbital overlaps are allowed by symmetry; the energies of the AOs are such that $1s$–$3s$ can occur as well as $1s$–$3p_z$; the diagram may also be explained in terms of the mixing of the $\sigma$ and non-bonding $3s$ levels in the MO diagram in question 1.

---

**Problem set 4.4**

1. Method as for worked example 4.1; $D(Cl–F) = 262$ kJ mol$^{-1}$. [Experimental value = 256 kJ mol$^{-1}$.]
2. Additivity gives $D(Cl–F) = 200.5$ kJ mol$^{-1}$. This method takes no account of the ionic contribution to the bonding.
3. Method as for worked example 4.1; $D(O–H) = 430$ kJ mol$^{-1}$.
4.

$$H_2O(g) \xrightarrow{2\bar{D}(O–H)} 2H(g) + O(g)$$

$\Delta_fH^\circ(H_2O, g)$     $2\Delta_aH^\circ(H) + \Delta_aH^\circ(O)$

$$H_2(g) + {}^1/_2O_2(g) \qquad \bar{D}(O–H) = 463.5 \text{ kJ mol}^{-1}$$

5. Variation in $Z_{eff}$ leads to variation in electron withdrawing power of atomic centre.
6. (a) From B to F (same quantum level for valence electrons), $Z_{eff}$ increases; (b) down the group, valence electrons further shielded from nuclear charge.

**Problem set 4.5**

1. (a) Scalar: magnitude; vector: magnitude and direction; (b) dipole moment is a vector.

2. Asymmetrical charge distribution; *in part* due to differing electronegativities of atoms. (*Care!* Read Section 4.13 in H&C: dipole moment of CO).

3. Polar: (b); (c); (e).     H–Cl    Cl–F    H–I

4. (a) H; (b) I; (c) Br; (d) I; (e) Rb; (f) Tl.

---

**Problem set 4.6**

1. B, $[He]2s^22p^1$; C, $[He]2s^22p^2$; N, $[He]2s^22p^3$; O, $[He]2s^22p^4$; F, $[He]2s^22p^5$; Ne, $[He]2s^22p^6$; Al, $[Ne]3s^23p^1$; Si, $[Ne]3s^23p^2$; P, $[Ne]3s^23p^3$; S, $[Ne]3s^23p^4$; Cl, $[Ne]3s^23p^5$; Ar, $[Ne]2s^22p^6$.

   (a) 3; (b) Ar; (c) Ne; (d) possibilities include Ne, $F^-$, $Na^+$, $Mg^{2+}$; (e) only with respect to its valence electrons; (f) 3– .

2. $K^+$

3. $O^{2-}$

4. Kr

5. (a) 20; (b) 14; (c) isoelectronic with $[S_2]^{2-}$, and with $[O_2]^{2-}$ and $Br_2$ with respect to the valence electrons.

6. Each is a 14 electron species.

7. (a) and (c) are isoelectronic with $CH_4$; (d) and (e) are isoelectronic with respect to the valence electrons.

---

**Problem set 4.7**

1.    :N : O:

2. •N=O ⟷ $\bar{N}$=$\overset{+}{O}$

3. (a) Bond order = 2; (b) paramagnetic.

➤ $\chi^P$ **values from Appendix 7 in H&C**

4. $\chi^P(N) = 3.0$; $\chi^P(O) = 3.4$, therefore polar but with O $\delta^-$. This contradicts the contribution made by one of the resonance forms.

5. N, $[He]2s^22p^3$; O, $[He]2s^22p^4$. Since $Z_{eff}$ O > N, the orbitals of O should be at lower energy than those of N.

6. See Figure 4.7.1.

**4.7.1** An approximate MO diagram for nitrogen monoxide assuming an LCAO approach (no orbital mixing).

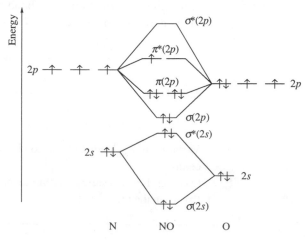

7.    (a) Paramagnetic; (b) bond order = 2.5; (c) $\sigma$ MO with %O 2s > %N 2s;
(d) $\pi^*$ MO with %N 2p > %O 2p.

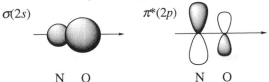

$\sigma(2s)$                    $\pi^*(2p)$

N    O                    N    O

8.    Differences: some MOs contains both s and p character; $\sigma$–$\pi$ crossover has
occurred. Explain in terms of mixing of the $\sigma^*(2s)$ and $\sigma(2p)$, labelled
according to Figure 4.7.1. Not easy to assess the bond order because the
nature of all the $\sigma$ MOs is not clear cut. Features unchanged: paramagnetism
and the fact the unpaired electron occupies a $\pi^*$ MO. The simple approach
gives a reasonable approximation to the bonding; the $\sigma$–$\pi$ crossover does
not influence the nature of the highest occupied MOs.

9.    (a) [NO]$^+$                                    (b)    NO → [NO]$^+$ + e$^-$
(c) Removal of electron from the $\pi^*$ MO.

---

**Problem set 4.8**

1.    :Cl:O·    or    ·Cl:O:
                                                    The left-hand structure is the
one you have probably drawn; this gives Cl an octet and makes the unpaired
electron oxygen-centred. The right-hand strucure is possible if the octet of
Cl is expanded; this gives Cl the odd electron. Without further information,
either structure is possible.

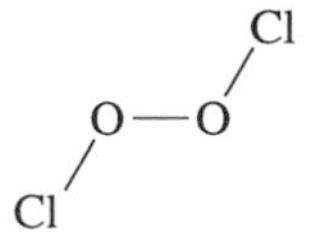

2.    See Figure 4.8.1.

3.    :Cl:O:
          :O:

**4.8.1**  Predicted structure of
$Cl_2O_2$. Note the similarity to
the structure of hydrogen
peroxide, Figure 3.1 in H&C.

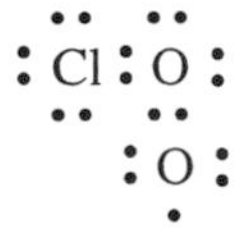

4.    Bond order is not clear cut: 1 or 2.
5.    O, [He]$2s^22p^4$; Cl, [Ne]$3s^23p^5$.
6.    See Figure 4.8.2.
7.    Bond order = 1.5; MO theory is in agreement with VB theory (see answer
to question 4).

**4.8.2**  An approximate MO
diagram for the formation of
ClO; an LCAO approach is
assumed.

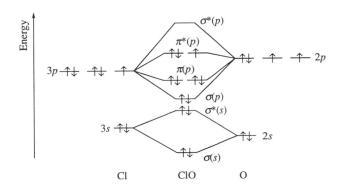

8.    (a) $\sigma(s)$ MO $\approx$ 50% Cl $3s$ and 50% O $2s$ character; (b) $\pi^*(p)$ MO $\approx$ 50% Cl $3p$ and 50% O $2p$ character. MO theory indicates that the unpaired elecron occupies an MO with both Cl and O character and is not localized on one centre.

---

**Problem set 5.1**

**5.1.1** The structures of $IF_5$, $SOF_4$ and $POCl_3$ predicted by VSEPR theory.

1.    (a) Tetrahedral; (b) trigonal planar; (c) octahedral; (d) trigonal bipyramidal; (e) see Figure 5.1.1a; (f) bent; (g) see Figure 5.1.1b; (h) see Figure 5.1.1c.

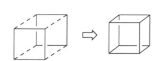

(a) Square-based pyramidal (based on an octahedron with one lone pair)

(b) Trigonal bipyramidal with the S=O double bond in the equatorial plane

(c) Tetrahedral

2.    (a) Square-based pyramidal (isoelectronic with $IF_5$, Figure 5.1.1); (b) linear; (c) tetrahedral; (d) tetrahedral; (e) octahedral; (f) tetrahedral; (g) tetrahedal; (h) trigonal pyramidal (based on a tetrahedon with one lone pair).

3.    (a) Less than 109.5° ($\approx$ 107.5°); (b) less than 109.5° ($\approx$ 104.5°); (c) 180°; (d) 90°; (e) 109.5°; (f) 109.5°.

---

**Problem set 5.2**

**5.2.1** A cube can be constructed from two squares in an eclipsed arrangement.

1.    All are predicted on the basis of *steric repulsions* between the groups attached to the central metal atom: (a) tetrahedral; (b) octahedral; (c) octahedral; (d) linear; (e) octahedral; (f) trigonal bipyramidal; (g) tetrahedral; (h) linear; (i) trigonal planar; (j) tetrahedral.

2.    No; a tetrahedral structure is expected on the basis of steric repulsions between the chlorine atoms. The Kepert model can never predict a square-planar structure.

3.    (a) Both a cube and a square-antiprism can be viewed as being fromed by two squares coming together (Figure 5.2.1 shows this for the cube); in the cube the squares are eclipsed and in the square-antiprism they are staggered. (b) A square-antiprismatic arrangement of cyano-groups is favoured because the inter-ligand distances are larger than in a cube, making the inter-ligand repulsions less.

---

**Problem set 5.3**

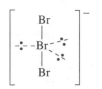

**5.3.1** The VSEPR model gives a linear shape for $[Br_3]^-$.

1.    Use the VSEPR model for both. O=C=O  No lone pairs on the carbon. In $SO_2$, the S atom has a lone pair.

2.    Use the VSEPR model for $[Br_3]^-$ (Figure 5.3.1) and Kepert for the $d$-block metal compound; Kepert model predicts a linear structure when there are two groups attached to the metal centre.

3.    Use the VSEPR model for both.  $SnCl_4$: Sn is in group 14; four bonding pairs of electrons, therefore tetrahedral.  $[ICl_4]^-$: I is in group 17; four bonding pairs and two lone pairs giving an arrangement of electron pairs based upon an octahedron; the ion is square-planar.

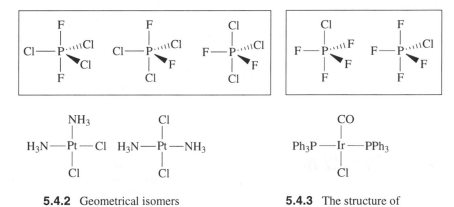

**5.3.2**  Structures of $SOCl_2$, FSN, $TeI_2Me_2$, $[SbF_5]^{2-}$ and $[BrO_4]^-$ predicted by the VSEPR model.

4.  Kepert model is used for $[FeCl_4]^{2-}$, $[Co(CO)_4]^-$ and $[IrCl_6]^{2-}$. Use VSEPR model for the others, summarized in Figure 5.3.2.

**Problem set 5.4**

1.  See H&C Section 5.11
2.  Shapes by VSEPR: (a) trigonal planar, no geometrical isomers; (b) trigonal pyramidal (based on a tetrahedral arrangement with one lone pair), no geometrical isomers; (c) trigonal bipyramidal, see Figure 5.4.1; (d) trigonal bipyramidal, see Figure 5.4.1; (e) octahedral, no geometrical isomers.
3.  By Kepert, $WCl_2F_4$ is octahedral. It has 2 isomers: *cis* (adjacent sites) and *trans* (opposite sites) determined with respect to the F atom positions.
4.  By Kepert, $WCl_3F_3$ is octahedral. It has 2 isomers: *mer* and *fac* (with respect to *either* the F or Cl atom positions). See Figure 5.22 in H&C.
5.  $[Ru(NH_3)_4Br_2]^+$ is octahedral (by Kepert) and has *cis* and *trans* isomers.
6.  See Figure 5.4.2.
7.  See Figure 5.4.3.

**5.4.1**  Geometrical isomers of $PCl_3F_2$ and $PClF_4$.

**5.4.2**  Geometrical isomers of $[Pt(NH_3)_2Cl_2]$.

**5.4.3**  The structure of Vaska's compound.

**Problem set 5.5**

1-2. The structures are drawn in Figure 5.5.1.
3.  (*E*) and (*Z*) describe the arrangements of groups with respect to a double bond. (*Z*) = 'on the same side of the bond' and (*E*) = 'on opposite sides of the bond' from the German words *zusammen* and *entgegen* respectively.
4.  Three isomers, two of which are geometrical (Figure 5.5.2).
5.  The diagram in Figure 5.7 (page 45) shows a linear environment at each N atom which is *incorrect*. Each N atom has a lone pair and the environment is bent. This leads to a series of geometrical isomers — how many?

**5.5.1** Structures for question 1, problem set 5.5. Only (b) possesses geometrical isomers.

**5.5.2** Isomers of $C_2H_2Cl_2$. We discuss *structural isomerism* in Section 8.4 of H&C.

Geometrical isomers

---

**Problem set 5.6**

1. Two (2 axial, 3 equatorial).
2. Follow the motions of atoms in Figure 5.6.1; this creates a new set of equatorial atoms (1, 2, 3) while atoms 4 and 5 become the new axial atoms. See H&C, Section 5.12. No P–F bonds are broken; all that happens is an angular deformation of the structure.
3. Berry pseudo-rotation. [Fe(CO)$_5$] also undergoes this process in solution.
4. Figure 5.8 (page 46) is *one step* in the process. A sequence of such steps eventually exchanges all the atom sites; e.g. take the right-hand structure in Figure 5.8; open the angle F(1)–P–F(2) to 180° whilst decreasing the angle F(4)–P–F(5) to 120°. Repeat the exercise with a different set of atoms. Repeat it again. Compare each result. Has every atom now 'visited' both an equatorial *and* an axial site? Does each F atom 'know' whether it is axial or equatorial?

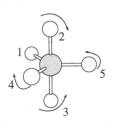

**5.6.1** The molecule PF$_5$ is stereochemically non-rigid (fluxional).

---

**Problem set 5.7**

1. (a) Linear but asymmetrical, polar; (b) trigonal pyramidal, lone pair on As atom, polar; (c) trigonal planar, non-polar; (d) tetrahedral, polar; (e) linear, symmetrical, non-polar; (f) bent, polar; (g) linear but asymmetrical, polar; (h) tetrahedral (Figure 1.3 of the *Self-Study Workbook*), non-polar.
2. See Figure 5.7.1: the N–F bond dipole moments in (E)-N$_2$F$_2$ cancel out, but they reinforce each other in (Z)-N$_2$F$_2$.
3. H$_2$O and OF$_2$ are bent molecules with two lone pairs on the O atom. The Pauling electronegativity values are $\chi^P(H) = 2.2$, $\chi^P(O) = 3.4$, $\chi^P(F) = 4.0$, meaning that the O–H bonds are polar in the sense $O^{\delta-}$—$H^{\delta+}$ and the O–F bonds are polar in the sense $O^{\delta+}$—$F^{\delta-}$. Now take into account dipole moments due to the presence of the lone pairs; their resultant will act in the *same* direction as the resultant of the O–H bond dipole moments, but will act in the *opposite* direction as the resultant of the O–F bond dipole moments. [Look at the discussion of NH$_3$ and NF$_3$ in the main text, Section 5.13.]

(E)-isomer

(Z)-isomer

**5.7.1** Bond dipoles in isomers of N$_2$F$_2$.

4. See Figure 5.7.2.

5. $O_2$ is a symmetrical diatomic so is non-polar. Figure 5.7.3 shows that $O_3$ is bent and contains two types of oxygen atom. (Draw resonance structures to describe the bonding in $O_3$.)

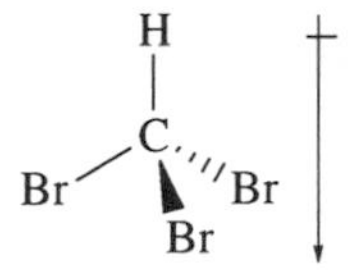

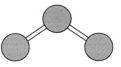

**5.7.2** $CHBr_3$: Tetrahedral structure and direction in which the resultant dipole moment acts.

**5.7.3** Ozone, $O_3$, is a bent molecule.

---

**Problem set 5.8**

1. Octet rule obeyed in (b), (c), (g), (h).

2. B has a sextet in $BCl_3$.

3. *p*-Block elements in the first row from B to F when they form compounds or ions; heavier *p*-block atoms *may* obey the octet rule; the octet corresponds to the completion of an $ns^2np^6$ (noble gas) configuration.

4. 18, corresponding to the completion of an $ns^2np^6nd^{10}$ electronic configuration; the expanded valence shell can accommodate between 8 and 18 electrons but the upper limit is rarely achieved (e.g. $[XeF_8]^{2-}$).

5. N has five valence electrons and forms three single bonds; in the ammonium ion, the centre to be considered is $N^+$ (isoelectronic with C) which has four valence electrons and can form four single bonds; $[NH_4]^+$ is isoelectronic with $CH_4$.

6. (a) One; (b) three; (c) four; (d) four; (e) two. [*Hint*: $O^-$ is isoelectronic with F; $O^+$ is isoelectronic with N; $B^-$ and $N^+$ are isoelectronic with C; $N^-$ is isoelectronic with O.]

---

**Problem set 5.9**

1. The covalent structure is the most important; a set of *three* ionic forms is needed to maintain equivalence of the three N–H bonds.

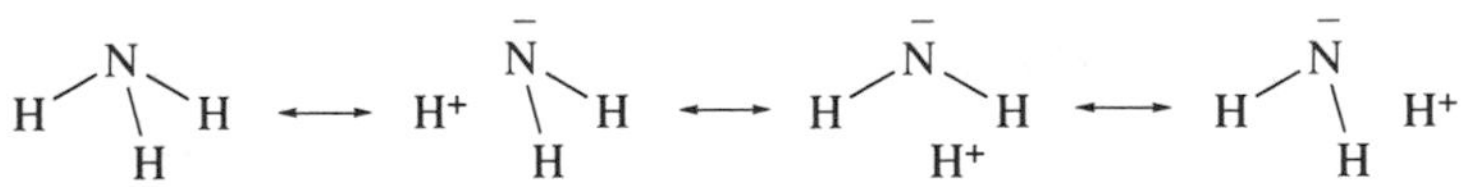

2.

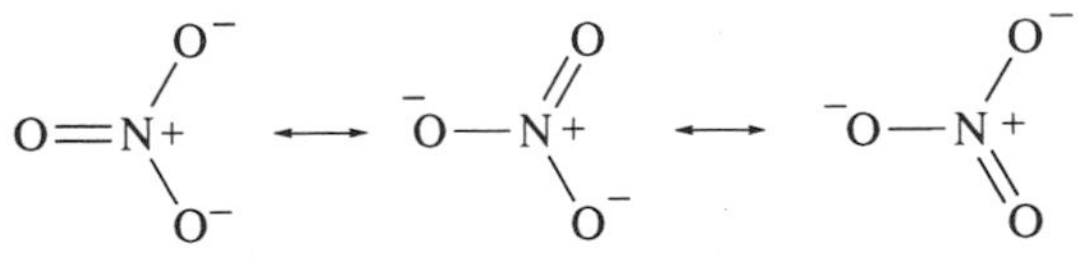

3.

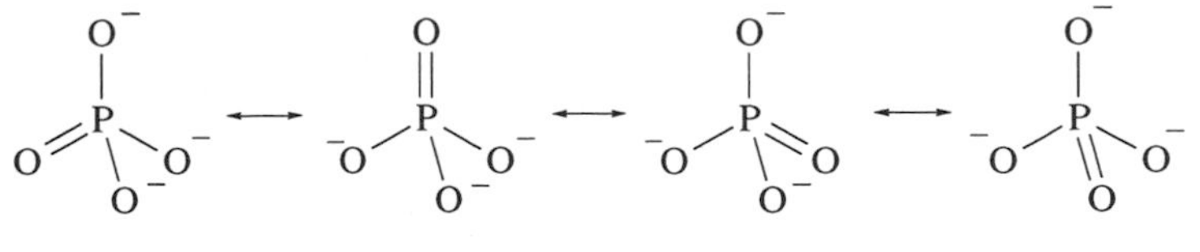

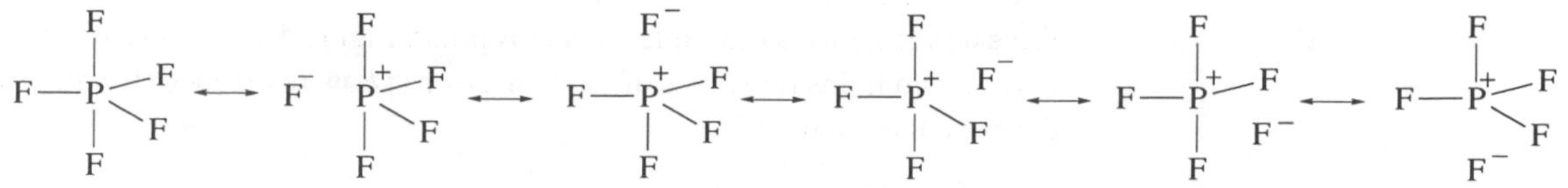

The resonance structures of $PF_6^-$ at the top of the page.

4.  See above.
5.  The resonance structure $\overset{-}{N}=\overset{+}{N}=\overset{-}{N}$ accounts for equivalence of N–N bonds.
6.  The resonance structure shown in Figure 5.9b (page 48) indicates that the valence shell of N has been expanded which cannot be the case: this structure is wrong. A set of resonance structures that does account for the planar structure with equivalent N–O bonds is:

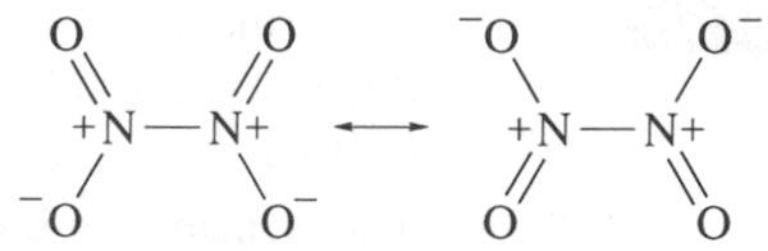

The resonance structures of $N_2O_4$.

---

**Problem set 5.10**

1.  (a) $sp^3$; (b) $sp$; (c) $sp^2$; (d) $sp$; (e) $sp^3$. Criteria: molecular shape and the numbers of bonding and lone pair domains.
2.  An $s$ and two $p$ AOs.
3.  For an atom using $ns$ and $np$ AOs, double bond formation requires the use of an unhybridized $np$ AO, which is 'left over' when $sp^2$ hybrids have been formed. The formation of $sp^3$ hybrid orbitals uses all three $np$ AOs.
4.  See Figure 5.10.1. The $[H_3O]^+$ ion is trigonal pyramidal with a lone pair. The lone pair and each bonding pair of electrons can be considered to occupy an $sp^3$ hybrid orbital; there are three *localized* O–H single ($\sigma$) bonds.
5.  $H_2CO$ is planar with a C=O double bond. An $sp^2$ hybridization scheme is appropriate (for the C–H and C–O $\sigma$-bonds), leaving a carbon $2p$ AO to overlap with an oxygen $2p$ AO to form the C–O $\pi$-bond.
6.  (a) Trigonal bipyramidal, $sp^3d$; (b) octahedral, $sp^3d^2$; (c) trigonal pyramidal with a lone pair, $sp^3$; (d) trigonal planar, $sp^2$; (e) square-based pyramidal with a lone pair, $sp^3d^2$; (f) T-shaped with 2 lone pairs, $sp^3d$.

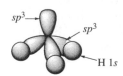

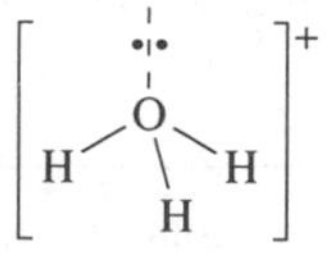

**5.10.1** An $sp^3$ hybridization scheme is appropriate for $[H_3O]^+$.

---

**Problem set 6.1**

1.  $IE_2$ (Mg, g) = $IE_1$(Mg$^+$, g)
2.  $IE_1$(Ne, g) > $IE_1$(Na, g). Removal of an electron from a filled quantum level in Ne requires more energy than the removal of the $3s^1$ electron from Na. Experimental values are 492 (Na) versus 2084 kJ mol$^{-1}$ (Ne).
3.  [He]$2s^1$ electron requires more energy to be removed than the [Ar]$4s^1$ electron (i.e. $IE_1$(Li, g) > $IE_1$(K, g); experimental values are 521 vs. 415 kJ mol$^{-1}$).
4.  Using the Hess cycle shown, $\Delta_rH$ = 1915 kJ per mole of reaction.

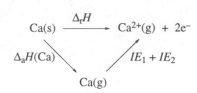

5. The overall enthalpy of reaction depends on several factors (see the Born-Haber cycle in Section 6.6 of the workbook); remember too that the *enthalpy* change is not the most reliable indicator of thermodynamic stability (see Chapter 12).

6. For each, look at the ground state electronic configuration of the atom: (a) H $1s^1$, He $1s^2$; removal of an unpaired electron versus one from a filled orbital; (b) He $1s^2$, Li $2s^1$; removal of an electron from a filled orbital versus an unpaired electron; (c) N $[He]2s^22p^3$, O $[He]2s^22p^4$; the half-filled $2p$ level of N has a particular stability associated with it making ionization more difficult than in O; (d) O $[He]2s^22p^4$, F $[He]2s^22p^5$, Ne $[He]2s^22p^6$; increase in nuclear charge and also sequential filling of the $2p$ level both mean that more energy is needed to remove an electron in going from O to F to Ne; (e) Li $2s^1$, Be $2s^2$; removal of an unpaired electron versus one from a filled $2s$ orbital; (f) Be $2s^2$, B $2s^22p^1$; removal of an electron from a filled $2s$ orbital versus the removal of an electron from the singly occupied, higher energy $2p$ orbital; (g) H $1s^1$, Li $2s^1$; both involve removal of an electron from a singly occupied $s$ AO, but the nuclear charge experienced by the electron in H is greater than that of the electron in Li where the core electrons shield the outer electron.

7. $IE_2 > IE_1$ because the second electron is being removed from a cation (greater effective nuclear charge); the larger difference between $IE_3$ and $IE_2$ than between $IE_2$ and $IE_1$ suggests that metal X is in group 2, and so the third electron is being removed from a filled $np$ level; values of $IE_4$ and $IE_5$ are influenced by stepwise increases in nuclear charge.

8. (a) $XCl_2$; (b) XO; (c) $X(OH)_2$.

---

**Problem set 6.2**

1. Negative enthalpy associated with attraction between the incoming electron and nucleus exceeds positive enthalpy associated with inter-electron repulsion.

2. In the reaction: $O^-(g) + e^- \rightarrow O^{2-}(g)$ there is significant repulsion between the incoming electron and the negative $O^-$ ion.

➤ **See question 4(b) in problem set 6.7**

3. The overall enthalpy change accompanying the formation of a metal oxide depends on several factors with the net effect that the endothermic reaction for the formation of $O^{2-}$ may be compensated.

4. Cl to I: electron added to $3p$ (Cl), $4p$ (Br) or $5p$ (I) orbital; the $5p$ AO is more diffuse than the $4p$ which is more diffuse than the $3p$ and so electron-electron repulsions are less as the electron enters the $5p$ than the $4p$ than the $3p$; hence, the reaction becomes *less* endothermic. F: although the electron-electron repulsions are greater as an electron enters the $2p$ than the $3p$ AO, the nucleus-electron attraction is sufficiently great to make the net process less endothermic than the trend for Cl to I might predict.

5. Using the Hess cycle shown, $\Delta_r H = 1812$ kJ per mole of reaction.

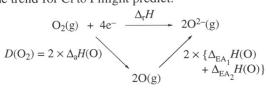

**Problem set 6.3**

1.  Smallest repeating unit in the lattice that unambiguously defines the lattice structure (see H&C, Section 6.7).

2.  The coordination number and geometry of $Cs^+$ are the same as those of $Cl^-$; either ion can occupy the central or corner site in the unit cell; the information given about the lattice as a whole is the same.

3.  The cell shown in Figure 6.5 (page 51) is not a *repeating unit*; try to place an identical cell of the type shown next to another so the corner ions are *shared* — what happens? Eight cells = smallest number that generate a repeating unit; see Figure 6.11 in H&C.

4.  In NaCl the cation : anion stoichiometry is 1:1; if $InBr_x$ had the identical structure, it too would have a 1:1 stoichiometry. However, if only one third of the cation sites are occupied, the stoichiometry must be $^1/_3 : 1$, i.e. 1 : 3. The formula is $InBr_3$ and indium is in oxidation state +3.

5.  (a) Formula is $CeO_2$; if the coordination number of O = 4, the stoichiometry dictates that the coordination number of Ce = 8; (b) given that the lattice type is one of those given in H&C, it must adopt the fluorite lattice.

6.  (a) GaP; (b) Ga in group 13 and P in group 15 are consistent with the formula; formally $Ga^{3+} P^{3-}$ although there will be significant covalent character.

7.  (a) (1 central Ir) + (8 corner Ir) = 2 Ir per unit cell; (4 O on faces) + (2 O completely inside the unit cell) = 4 O; stoichiometry is 2 : 4, i.e. 1 : 2; (b) this is the rutile lattice-type, but the diagram shows a different orientation than the one given in Figure 6.16 in H&C.

---

**Problem set 6.4**

1.  (a) O, 4; Nb, 4; (b) square planar; (c) NbO; (d) NaCl lattice-type in which the corner $Na^+$ and central $Cl^-$ sites are vacant.

2.  (a) NiAs; FeSn; (b) 6; (c) trigonal prismatic.

3.  Fluorite is $CaF_2$, with eight-coordinate $Ca^{2+}$ ions and four-coordinate $F^-$ ions. If potassium oxide adopted the same lattice structure, the formula would be $KO_2$; in order to make the stoichiometry 2 : 1 instead of 1 : 2, exchange the cation and anion positions; similarly for potassium sulfide. An *antifluorite* lattice-type is one in which the cation and anion occupancies are the reverse of those in a fluorite lattice.

4.  (a) see Figure 6.4.1; (b) it requires three rutile-type unit cells in order that the infinite lattice of $FeSb_2O_6$ can be described; (c) O: 3-coordinate (trigonal planar); Fe: 6-coordinate (octahedral); Sb: 6-coordinate (octahedral); (d) Fe: one central + eight corners = 2 Fe; Sb: two central + eight edge = 4 Sb; O: ten central + four face = 12 O; stoichiometry = 2 : 4 : 12 = 1 : 2 : 6; (e) assuming that oxygen is $O^{2-}$, there are three possibilities for iron and antimony: Fe(II) and Sb(V), or Fe(VI) and Sb(III), or Fe(IV) with mixed Sb(III) and Sb(V), and of these the first is the most likely.

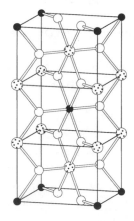

**6.4.1** A unit cell of the trirutile lattice adopted by $FeSb_2O_6$. The compound $ZnSb_2O_6$ also adopts this structure; which sites would the zinc centres occupy?

| **Problem set 6.5** | 1. | Using X-ray or neutron diffraction methods and analysing the data as described in Section 6.14 of H&C. |
|---|---|---|
| | 2. | (a) With the data set provided, it is not possible to obtain a value of $r_{K^+}$; (b) given the value of $r_{Na^+}$, you could find $r_{F^-}$ and hence $r_{K^+}$, but a better estimate can be made by using more of the data and generating a *self-consistent* set of radii. The tabulated value of $r_{K^+}$ is 138 pm. |
| | 3. | Increase. |
| | 4. | On going from O to $O^{2-}$ the radius increases as the effective attraction between the nucleus and the electrons decreases. |
| | 5. | On going from Ca to $Ca^{2+}$ the radius decreases as the effective nuclear charge increases. |
| | 6. | Effective nuclear charge for $Mn^{4+} > Mn^{3+} > Mn^{2+}$. |

**Problem set 6.6**    1.

➤

**Each constant or variable is defined in H&C, Section 6.15**

$$\text{Lattice energy} = \Delta U(0\text{ K}) = -\left( \frac{L \times A \times |z_+| \times |z_-| \times e^2}{4 \times \pi \times \varepsilon_0 \times r} \right) \times \left( 1 - \frac{1}{n} \right)$$

(a) the first term and (b) the second term on the right-hand side of the equation; see H&C, Section 6.15.

2. Internal energy, $\Delta U(0\text{ K})$.
3. (a) 8.5; (b) 7; (c) 7; (d) 9; (e) 8.
4. $\Delta U(0\text{ K}) = -759.5$ kJ mol$^{-1}$

| **Problem set 6.7** | 1. | $Mg^{2+}(g) + 2Cl^-(g) \rightarrow MgCl_2(s)$    (Pay attention to states). |
|---|---|---|
| | 2. | See Figure 6.7.1 |
| | 3. | An enthalpy change; $\Delta U(0\text{ K}) \approx \Delta H$ (298 K) |
| | 4. | For each, use a Born-Haber cycle similar to that in Figure 6.7.1. (a) $CaF_2$ $\Delta_{lattice}H^\circ = -2645$ kJ mol$^{-1}$; (b) $K_2O$ $\Delta_{lattice}H^\circ = -2277.5$ kJ mol$^{-1}$ |

**6.7.1** A Born-Haber cycle for the estimation of the lattice energy of magnesium chloride. The result is an associated *enthalpy* value.

$$
\begin{array}{ccc}
Mg(s) + Cl_2(g) & \xrightarrow[\text{where } D(Cl_2) = 2 \times \Delta_a H^\circ(Cl)]{\Delta_a H^\circ(Mg) + D(Cl_2)} & Mg(g) \quad + \quad 2Cl(g) \\
\Delta_f H^\circ(MgCl_2, s) \downarrow & & \downarrow IE_1 + \qquad \qquad \downarrow 2 \times \Delta_{EA}H(Cl) \\
& & IE_2 (Mg) \\
MgCl_2(s) & \xleftarrow{\Delta_{lattice}H^\circ(MgCl_2, s)} & Mg^{2+}(g) \quad + \quad 2Cl^-(g)
\end{array}
$$

| **Problem set 7.1** | 1. | (a) Simple cubic, 6 nearest neighbours; (b) body-centred cubic, 8 nearest neighbours. |
|---|---|---|
| | 2. | Less efficient. |
| | 3. | (a) ABABAB…; (b) ABCABC… |
| | 4. | 12 nearest neighbours in each. |
| | 5. | (a) See Figure 7.1.1; the other set of three holes is equally valid; (b) no; (c) see Figure 7.1.1. |

For part (a)

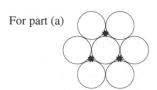

For part (c)

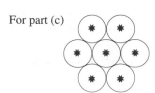

**7.1.1** Packing of spheres: question 5.

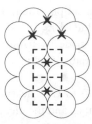

✱ = octahedral hole
✘ = tetrahedral hole

**7.1.2** Packing of spheres: question 6.

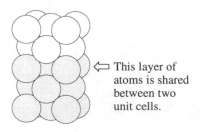

⇐ This layer of atoms is shared between two unit cells.

**7.1.3** Packing of spheres: question 7.

6.    (a) See Figure 7.1.2; (b) cubic close-packing; (c) octahedral and tetrahedral; (d) see Figure 7.1.2, the holes are completed when the lattice is extended.

7.    (a) Cubic close-packing; (b) 2; (c) see Figure 7.1.3; (d) face-centred cubic (fcc).

8.    It is an ABA… arrangement and so is present only in (b) hcp.

---

**Problem set 7.2**

1.    By using Pythagoras's theorem:   ratio $r_{X^-} : r_{M^+} = 2.414$.

2.    Determine the ratio $r_{X^-} : r_{M^+}$ for each compound: LiF 1.75; NaF 1.30; KF 0.96; CsF 0.78; NaCl 1.77; KCl 1.31; LiBr 2.58; LiI 2.89. Only LiBr comes close to having the ideal anion:cation radius ratio. In LiI, the holes are too large for the cations. In LiF, NaF, NaCl and KCl, the relative sizes of the cations mean that the anions cannot touch each other and are not close-packed. In CsF and KF, the cations are larger than the anions (see below).

3.    CsF and KF are the 'odd ones out' because the cations are larger than the anions. Since the NaCl lattice has identical environments for each cation and anion, CsF and KF can still be classed as having the NaCl-type lattice.

---

**Problem set 7.3**

➤

**More detailed discussion of the structures of these elements in the solid state is given in Chapter 7 of H&C**

1.    (a) Cu has a fcc lattice (Table 7.3, H&C) and on melting, the lattice collapses but there are still significant Cu···Cu interactions; (b) solid hydrogen consists of $H_2$ molecules in an hcp arrangement and on melting the *intermolecular* interactions are essentially destroyed; (c) solid bromine consists of $Br_2$ molecules interacting to form zigzag chains organized in layers; melting destroys most *intermolecular* interactions; (d) in the solid state, $S_8$ molecules (rings) are organized in an assembly with van der Waals interactions between the rings; on melting, most *intermolecular* interactions are lost; (e) argon is monatomic and in the solid the atoms are in a ccp array; during the solid to liquid transition, most *interatomic* interactions are destroyed (the liquid range is *very* short); (f) sodium possesses a bcc lattice and on melting the regular array of atoms is lost but significant Na···Na interactions remain.
Values of mp [Cu 1358; $H_2$ 13.7; $Br_2$ 266; $S_8$ 388; Ar 84; Na 371 K] can be

rationalized in terms of the types of interactions between atoms or molecules in the solid state; the highest value for Cu represents loss of strong metal-metal bonding interactions (these are *not* localized Cu–Cu bonds, see Section 7.5); sodium is metallic but the atoms are not close-packed as in copper.

2. (a) $Cl_2(g) \rightarrow 2Cl(g)$; (b) $Zn(s) \rightarrow Zn(g)$; (c) $P_4(s) \rightarrow 4P(g)$; (d) $K(s) \rightarrow K(g)$.

3. $D(F–F) = 2 \times \Delta_a H^\circ$

4. (a) Group 1 metals possess bcc lattices with less efficiently packed atoms than most of the *d*-block metals; (b) an exception is Hg which has a distorted cubic lattice; Hg is a liquid at 298 K.

> **Another exception is Mn: see Section 7.11 in H&C**

5. B and Al are first and second elements in group 13; β-rhombohedral B has an infinite lattice (Figure 7.21 in H&C) with covalent B–B bonds, most of which must be broken upon melting; Al is metallic and possesses a ccp lattice. It requires less energy to rupture the metal-metal bonding than to break up the extensive covalent network in β-rhombohedral B.

6. (a) $N_2(s) \rightarrow N_2(l)$; (b) $N_2(l) \rightarrow N_2(g)$; $N_2(g) \rightarrow 2N(g)$; both (a) and (b) are tabulated per mole of $N_2$ molecules, but (c) is per mole of N atoms.

7. (a) All are diatomic molecules; in the solid state only weak van der Waals forces operate between the molecules; on melting most intermolecular interactions are destroyed and it takes little more energy to overcome the remainder i.e. to reach the boiling point. (b) Rationalize in terms of solid state structure and elemental electronic structure, recalling that for each element the enthalpy change concerns the change from the element in its standard state to gaseous atoms, and is quoted *per mole of atoms*: the metals Li (bcc) and Be (hcp) undergo loss of metal-metal bonding and the hcp lattice requires more energy to disrupt than the bcc; also, Be has more valence electrons than Li; B and C both possess infinite covalent lattices which must be completely ruptured to form gaseous atoms; for N, O and F, atomization involves cleaving diatomic molecules, an N≡N triple bond, an O=O double bond and an F–F single bond. (c) $C(s) \rightarrow C(l)$ involves breaking covalent bonds; for nitrogen, the transition from solid to liquid involves only the loss of van der Waals interactions. (d) Factors as in part (b).

> **Relationship of valence electrons to metal-metal bonding: see Section 7.13 in H&C**

8. Density depends upon the number of carbon atoms in a given volume; both diamond and α-graphite have infinite covalent lattices but the atoms in diamond are more effectively packed giving a higher density.

9. $C_{60}$ and $C_{70}$ are molecular; diamond and α-graphite have infinite covalent lattices. Separation by column chromatography requires that the components to be separated are soluble in some solvent or solvent mixture; $C_{60}$ and $C_{70}$ are soluble in hexane and benzene; hexane can be used to elute these allotropes.

> **See Figure 7.18 in H&C: $C_{60} \cdot 4C_6H_6$ is obtained by recrystallizing $C_{60}$ from benzene**

**Problem set 7.4**

1. 2.6 A

2. $6 \times 10^{-8} \, \Omega \, m$

3. $0.27 \, \Omega$

4. 22.2 A; increasing the cross section of the wire lowers the resistance allowing a greater current to flow.

5. See Figure 7.23 in H&C for structure of α-graphite; electrons in the plane

of the fused hexagonal rings allow current to flow in this plane only.
6.    (a) ≈ $3.3 \times 10^{-5}$ Ω; (b) ≈ $6.2 \times 10^{-5}$ Ω

---

**Problem set 7.5**

1-2.   See Section 7.13 in H&C.
3.    (a) (iii); (b) (ii); (c) (i)
4.    Si group 14, As group 15; substituting As for Si atoms increases the number of valence electrons; this provides more electrons in a band close to the conduction band and thermal population of the latter is more readily achieved than in pure silicon. The increased conductivity can be described in terms of the extra electron (from As) migrating from As to Si to Si ... atoms as the electron is excited from a band with As orbital character to a band with Si orbital character.

---

**Problem set 8.1**

1- 2.   Refer to Figure 8.1 and Section 8.1 in H&C.
3.    (a) C=C; (b) C≡C; unsaturated.
4.

| | | | |
|---|---|---|---|
| $sp^3$ $sp^3$ $sp^3$ $sp^3$ | $sp^3$ $sp^2$ $sp$ $sp$ $sp^2$ | $sp^2$  $sp^2$ $sp^3$ | $sp^3$ $sp$ $sp$ $sp^3$ |
| $CH_3CH_2CH_2CH_3$ | $CH_3CH=C=C=CH_2$ | $CH_2=CHCH_3$ | $CH_3C≡CCH_3$ |

---

**Problem set 8.2**

1.    $C_nH_{2n+2}$
2.    (a) $C_4H_{10}$; (b) $C_8H_{18}$; (c) $C_{10}H_{22}$; (d) $C_6H_{14}$
3.    (a) 2,2-dimethylpropane; (b) 2,4-dimethylpentane; (c) 2,2-dimethylpentane; (d) 3-ethyl-5-methylheptane.
4.    (a)                          (b)                          (c)

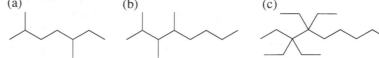

5.    Three:

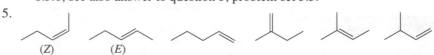

6.    (a) $(CH_3)_3C$ or $Me_3C$; (b) $(CH_3)_2CH$ or $Me_2CH$

---

**Problem set 8.3**

1.    $C_nH_{2n}$
2.    (a) $CH_3CH_2CH=CH_2$; (b) $CH_3CH_2CH=CHCH_3$; (c) $CH_2=CHCH=CH_2$; (d) $CH_3CH=CHCH=CHCH_3$
3.    (a) Hex-1-ene; (b) buta-1,2-diene; (c) hex-3-ene.
4.    Used to distinguish between geometrical isomers of an alkene; see Figure 8.3.1; see also answer to question 3, problem set 5.5.
5.

(E)-isomer    (Z)-isomer

**8.3.1**  Geometrical isomers of but-2-ene.

          (Z)          (E)

---

**Problem set 8.4**

1.    Linear
2.    (a) ══════  (b)           (c)

**8.5.1**  Answer to question 1, problem set 8.5.

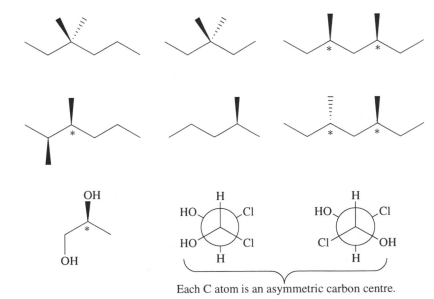

3.  Yes — the triple bond may be in one of two positions: $CH_3CH_2C{\equiv}CH$ or $CH_3C{\equiv}CCH_3$.
4.  No.

---

**Problem set 8.5**

1.  See Figure 8.5.1
2.  (a) Staggered; (b) skew; (c) eclipsed; (d) eclipsed.
3.  (a) See Figure 8.5.2; (b) rotation is about a *terminal* C–C bond; (c) comparing structures (a) to (d), the only changes are the relative positions of the H atoms attached to atoms C(1) and C(4), and C(1) and C(2) [numbering from the right] — the closer the non-bonded H atoms approach, the higher (less negative) the steric energy of the molecule becomes.
4.  Conformers: (a) no; (b) yes; (c) yes; (d) no; rotation is possible about *single* bonds present in the molecules, but rotation about C–H bonds will not alter the conformation.
5.  Extended.

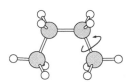

**8.5.2**  Answer to question 3(a), problem set 8.5.

---

**Problem set 8.6**

1.  Asymmetric carbon atom is designated by * :

Each C atom is an asymmetric carbon centre.

2.  (a) Mirror plane through C(3), passing between the two methyl C atoms; (b) mirror plane through C(3), passing between the two methyl C atoms; (c) mirror plane through C(4) perpendicular to plane of the paper; (d) no mirror plane; (e) mirror plane passing between Me groups and containing C(2)–H bond; (f) no mirror plane; (g) no mirror plane; (h) no mirror plane; (i) mirror plane bisecting the C–C bond. If you cannot see why a molecule contains a plane of symmetry, consider the consequences of free rotation. Stereochemical formulae are essential for (c), (f), (h) and (i), (why?).

3.  Chiral: (d), (f), (g), (h). In compounds (c) and (i), there are *two* chiral atoms *and* a plane of symmetry. These molecules are achiral.

4.  Each chain is helical; may have a right- of left-handed twist; such pairs of chains are non-superimposable mirror images and so are enantiomers.

---

**Problem set 8.7**

1.  −5472 kJ mol⁻¹
2.  599 dm³ ≈ 0.6 m³
3.  $C_6H_{14}$

4.  −126 kJ mol⁻¹

---

**Problem set 8.8**

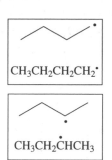

CH₃CH₂CH₂CH₂•

CH₃CH₂ĊHCH₃

**8.8.1** Radicals formed by the abstraction of H• from butane.

1.  (a) HBr, $CH_3Br$, $CH_2Br_2$, $CHBr_3$, $CBr_4$ plus products of chain growth; e.g. $C_2H_6$; (b) irradiation initiates the reaction: $Br_2 \xrightarrow{h\nu} 2Br^\bullet$

2.  (a) Atoms C(1) and C(4) are primary centres; C(2) and C(3) are secondary centres; (b) see Figure 8.8.1; (c) no, abstraction of H• to give a radical with a secondary carbon centre is preferable to abstraction to give a radical with a primary carbon centre.

3.  (a) $Br_2 \xrightarrow{h\nu} 2Br^\bullet$

    (b) $CH_3CH(Me)CH_3 + Br^\bullet \rightarrow HBr + CH_3CH(Me)CH_2^\bullet$

    $CH_3CH(Me)CH_3 + Br^\bullet \rightarrow HBr + CH_3\dot{C}(Me)CH_3$

    Further propagation steps lead to additional hydrogen abstraction.

    (c) The second product is favoured; the radical formed possesses a tertiary carbon centre.

    (d) $CH_3CH(Me)CH_2^\bullet + Br^\bullet \rightarrow CH_3CH(Me)CH_2Br$

    $CH_3\dot{C}(Me)CH_3 + Br^\bullet \rightarrow CH_3C(Me)(Br)CH_3$

    $2Br^\bullet \rightarrow Br_2$

4.  Reaction between $CH_3^\bullet$ and HBr leads to $CH_4$ and $Br^\bullet$; this is the reverse of a propagation step and reforms reactants.

5.  $C_2H_5^\bullet$ is formed in a propagation step; reaction between these radicals leads to chain growth: $2C_2H_5^\bullet \rightarrow C_4H_{10}$

---

**Problem set 8.9**

1.  (a) $Br_2$; (b) HBr; (c) $H_2$ with catalyst (e.g. Ni, Pd); (d) alkaline, aqueous $KMnO_4$; (e) HCl; (f) $H_2O$ with acid catalyst; (g) hydroboration followed by reaction with $H_2O_2$. Polar solvents facilitate electrophilic additions.

2.  Decoloration of the initially orange solution; addition of $Br_2$ to ethene gives 1,2-dibromoethane removing the orange colour (i.e. due to $Br_2$).

3.

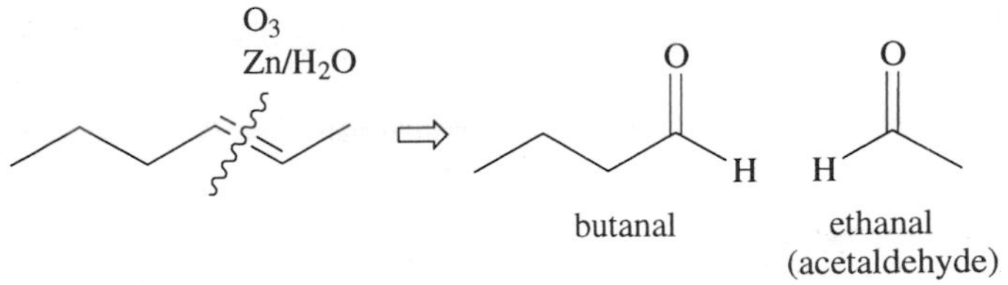

O₃
Zn/H₂O

butanal

ethanal
(acetaldehyde)

4.  2-Chloro-2-methylpropane.

5.  An organic dichloride with the two chlorine atoms attached to *adjacent* carbon atoms.

6.  This is double bond migration and is catalysed by either acid or base.

---

**Problem set 8.10**

1.  Region of electron density in the C=C double bond; $\pi$-electrons are involved in bond formation with an electrophile.

2.  $\pi$-Electrons induce a dipole in the initially non-polar $Cl_2$ bond.

3.  (a)

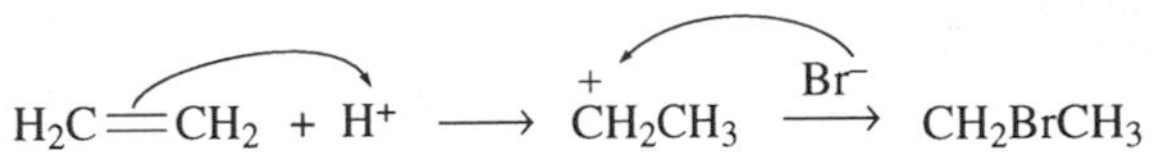

$$H_2C=CH_2 + H^+ \longrightarrow \overset{+}{CH_2}CH_3 \overset{Br^-}{\longrightarrow} CH_2BrCH_3$$

(b) See Figure 8.10.1; ensure the first step has the larger activation energy; (c) the first step; i.e. the formation of the intermediate.

4.  Preferential attachment of a hydrogen atom from HX to the carbon atom in the alkene that bears the most H atoms (e.g. H attaches to a primary rather than secondary C centre):

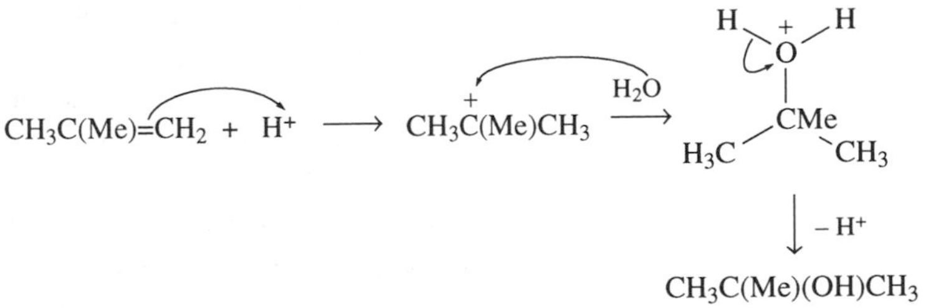

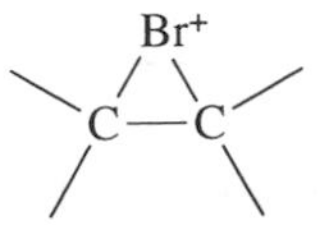

$$CH_3C(Me)=CH_2 + H^+ \longrightarrow CH_3\overset{+}{C}(Me)CH_3 \overset{H_2O}{\longrightarrow}$$

$$\downarrow -H^+$$

$$CH_3C(Me)(OH)CH_3$$

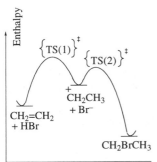

**8.10.1** Reaction profile for the addition of HBr to ethene.

**8.10.2** Bromonium ion intermediate.

5.  (a) Secondary; (b) primary; (c) secondary; (d) tertiary; (e) primary; (f) tertiary.

6.  (c) > (b) ≈ (d) ≈ (e) > (a)

7.  The addition proceeds predominantly to form $CH_3CHClCH_3$ even though it could also form $CH_3CH_2CH_2Cl$.

8.  Intermediate in the chlorination is a carbenium ion, but in the bromination it is a bromonium ion (Figure 8.10.2); bromination gives *anti*-addition products indicating that the intermediate is such that it blocks addition of $Br^-$ from the same side as initial attack by $Br^+$.

**8.11.1** Sawhorse drawings of *erythro-* and *threo-*pairs of enantiomers of ABHC–CABX.

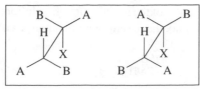

*erythro-*pair                    *threo-*pair

---

**Problem set 8.11**

1.  See Figure 8.11.1
2.  (a)
    $$MeHC \!=\! CHMe + Cl\!-\!Cl \rightarrow MeHClC\!-\!\overset{+}{C}HMe \rightarrow MeCHClCHClMe$$
    with Cl$^-$ attacking
    (b) yes;  (c) three (why?).
3.  The intermediate is a bromonium rather than a carbenium ion; this blocks one side for attack by Br$^-$ and allows only *anti*-addition. There will be one pair of enantiomers (i.e. two stereoisomers).
    [You can confirm the answers concerning stereoisomers in questions 2(c) and 3 by substituting Cl or Br for H and X; see Figure 8.11.1.]

---

**Problem set 8.12**

1.  (a) Radical initiator; (b) a primary radical (e.g. equation 8.55 in H&C) may rearrange to a more stable secondary radical:
    $$RCH_2CH_2^{\bullet} \rightarrow R\overset{\bullet}{C}HCH_3$$
    Participation of the secondary radical in the chain growth results in the introduction of a methyl branch; (c) less efficient packing for the branched chain; (d) less efficient packing means fewer molecules per unit volume, i.e. a lower density (see next answer).
2.  The high-density polymer suggests linear chains (no, or perhaps few, branches) which pack efficiently in the solid state.
3.  Melting point is dependent upon the van der Waals interactions between the polymer chains; these are weaker in the case of the branched chains and the low-density polyethene will have a lower melting point than the high-density form.
4.  (a) Isotactic polymer is *stereoregular* with the side-chains (Me groups in polypropene) arranged on the same side of the carbon backbone; (b) enhanced crystallinity achieved for the stereoregular polymers giving better physical properties (e.g. rigidity, higher melting point) for commercial purposes.

---

**Problem set 8.13**

➤ **Values of bond dissociation enthalpies: see Section 3.5 in H&C**

1.  $-1937.5$ kJ mol$^{-1}$
2.  (a) $-165$ kJ mol$^{-1}$; (b) $-125$ kJ mol$^{-1}$; the difference of 40 kJ can be rationalized as the difference between the bond dissociation enthalpies of C≡C and C=C, and C=C and C–C. [Consider the hydrogenations of propyne and propene in terms of bonds broken and made.]
3.  Each addition step follows Markovnikov's pattern.
4.  (a) $CH_3CBr_2CHBr_2$; (b) $CH_3CH_2CH_2CH_2CH_2CH_3$; (c) $CH_3CCl_2CMe_2CH_3$.
5.  $CH_3CH_2CH_2C\equiv CH \rightarrow CH_3CH_2CH_2CBr=CH_2 \rightarrow CH_3CH_2CH_2CBrICH_3$.
6.  Only terminal alkynes possess a C≡C–H unit; for acidic behaviour, we require

the equilibrium:

$$RC{\equiv}CH + H_2O \rightleftharpoons [RC{\equiv}C]^- + [H_3O]^+$$

7. (a) $Na^+[CH_3CH_2C{\equiv}C]^- + \frac{1}{2}H_2$; (b) $Na^+[CH_3C{\equiv}C]^- + NH_3$; (c) no reaction.

8. $Na[C{\equiv}CH] + H_2O \rightarrow NaOH + HC{\equiv}CH$

9. (a) $CaC_2$ or $Ca^{2+}[C_2]^{2-}$; (b) $CaC_2 + 2H_2O \rightarrow Ca(OH)_2 + HC{\equiv}CH$; (c) look at

Figure 6.10 on page 55 ➤

Figure 6.10 in the workbook — in $CaC_2$, the $Ca^{2+}$ ions occupy one set of octahedral sites (e.g. those occupied by $Na^+$ in Figure 6.10) and the $[C_2]^{2-}$ ions occupy the second set of sites. However, the $[C_2]^{2-}$ ions are not spherical; taking Figure 6.10 as your starting point, place each $[C_2]^{2-}$ so the midpoint of the C≡C bond lies at the site occupied by a chloride anion in Figure 6.10 with the C–C vectors all directed vertically; the result is that the unit cell of $CaC_2$ is elongated along one axis compared to the NaCl unit cell.

---

**Problem set 9.1**

1. $9 \times 10^{-4}$ mol dm$^{-3}$

2. See Figure 9.1.1; the gradient gives $\varepsilon = 211$ dm$^3$ mol$^{-1}$ cm$^{-1}$; a series of readings allows for errors on individual readings — not all the values are accurate (see Figure 9.1.1).

3. (a) $9.6 \times 10^{-4}$ mol dm$^{-3}$; (b) 0.043 g

4. 0.347 M

5. 0.0397 g; how great an error is there in the absorbance if you use 0.04 g?

6. (a) Before the flash, **9.8** absorbs a constant amount of light at $\lambda = 435$ nm. At the point of the flash the absorbance drops to a minimum value as **9.8** isomerizes; $\lambda_{max}$ for the (Z)-isomer is *not* at 435 nm. After the flash, isomerization occurs slowly to reform compound **9.8**. (b) 16 667 dm$^3$ mol$^{-1}$ cm$^{-1}$.

**9.1.1** Plot of absorbance against concentration using the data in question 2, problem set 9.1. Since the path length is 1 cm, the gradient gives the extinction coefficient directly.

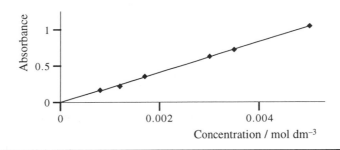

---

**Problem set 9.2**

1. (a) $1.14 \times 10^{-26}$; (b) $1.53 \times 10^{-27}$; (c) $1.61 \times 10^{-27}$; (d) $1.11 \times 10^{-27}$ kg.

2. 2659 cm$^{-1}$

3. 1878 N m$^{-1}$

4. The trend in values of bond dissociation enthalpies should approximately follow that of force constants.

5. 2735 cm$^{-1}$

6. Shift in $\bar{\nu} = 741$ cm$^{-1}$ to lower wavenumber ($\bar{\nu} = 2042$ cm$^{-1}$).

7. Changing the CN group from (a) $(^{12}C)(^{14}N)$ to $(^{13}C)(^{14}N)$, or (b) from $(^{12}C)(^{14}N)$ to $(^{12}C)(^{15}N)$ results in a *small* change in the reduced mass. If we assume that the force constants of each bond are ≈equal (a valid assumption), then we expect only a small shift in the IR spectroscopic absorption. From

Figure 9.2 on page 78 ➤

Figure 9.2, $\bar{\nu} \approx 2250$ cm$^{-1}$. After labelling, $\bar{\nu} =$ (a) 2203, (b) 2215 cm$^{-1}$.

1.  (a) $CH_3CH_2OH$ or ⟍⟋⟍OH ; (b) O–H stretching mode; absorption is broad due to hydrogen bonding; (c) stretching of C–H bonds; (d) bands due to C–H and O–H stretches should be similar, but the fingerprint regions should be different, diagnostic of each compound.

2.  $HC≡CH$ is a symmetrical molecule (same groups attached to the $C≡C$ carbon atoms; vibration of the $C≡C$ bond does not lead to a change in the molecular dipole moment and the mode is IR inactive. Vibration of the $C≡C$ bond in $CH_3C≡CH$ (propyne) does result in a change in the molecular dipole moment and the mode is IR active.

3.  The C–H stretch occurs at 3311 $cm^{-1}$ ($sp$ hybridized carbon), the $C≡N$ stretch at 2097 $cm^{-1}$ and the bend (Figure 9.3.1) at 712 $cm^{-1}$.

4.  In octane, the absorptions near 3000 $cm^{-1}$ are the C–H stretches; other bands lie in the fingerprint region. In oct-1-ene, bands near 3000 $cm^{-1}$ are the C–H stretches, and the weaker absorption near 1600 $cm^{-1}$ is the $C=C$ stretch; other bands lie in the fingerprint region. In methyl octanoate, bands near 3000 $cm^{-1}$ are the C–H stretches, and the strong absorption at 1745 $cm^{-1}$ is assigned to the $C=O$ stretch; other bands lie in the fingerprint region.

5.  (a) By VSEPR, $CO_2$ is linear and $SO_2$ is bent:

$$O=C=O \qquad \overset{..}{\underset{O \diagdown \diagup O}{S}}$$

(b) The number of degrees of vibrational freedom for $CO_2$ = 4 but two are degenerate and the symmetric stretch is IR inactive; two absorptions should be observed. For $SO_2$, the number of degrees of vibrational freedom = 3. (c) 2349 $cm^{-1}$ is the asymmetric stretch of $CO_2$.

6.  Formation of two aldehydes occurs:

oct-1-ene                    heptanal        methanal

Each $C=O$ stretch gives rise to a strong absorption ≈ 1700 $cm^{-1}$.

7.  (a) See Figure 9.3.2; (b) a C–Cl stretch is expected in the approximate range 600-800 $cm^{-1}$. The strong absorption at 573 $cm^{-1}$ may be reasonably assigned to the C–Cl stretch but the weaker band at 810 $cm^{-1}$ possibly makes the assignment ambiguous. Absorptions due to C–X vibrations (X = halogen) are not always readily observed and assigned as is seen for 1-chlorooctadecane (Figure 9.3.4).

8.  (a) Trigonal planar (use VSEPR); (b) S=O bond stretch *together* (as shown in Figure 9.3.3) and then are compressed *together* ; (c) IR inactive; no change in molecular dipole moment is produced.

9.  The question does not specify the type of derivative, so do *not* assume that it is an addition product. The most notable feature in the spectrum is the lack of absorptions due to C–H stretches — the compound is fully chlorinated (*perchlorated*). The band near 1600 $cm^{-1}$ suggests C=C double bond character. The spectrum is actually that of 1,1,2,3,4,4-hexachlorobuta-1,3-diene.

bend
(deformation)

**9.3.1** The bending mode of vibration of the HCN molecule.

➤
**See H&C, page 411**

➤ **See H&C, Section 8.13**

**9.3.2** Structure for question 7, part (a).

**9.3.3** The symmetric stretch of $SO_3$.

**9.3.4**  The IR spectrum of 1-chlorooctadecane.

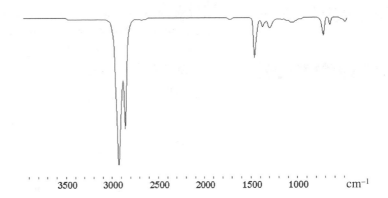

10.  The sharp bands at 2500 cm$^{-1}$ are due to C≡N stretches. Apart from the C≡N carbon, compound **I** has only $sp^2$ C centres but **II** has both $sp^2$ and $sp^3$ C centres. This affects the stretching frequencies of the C–H bonds. Spectrum **A** shows C–H stretches at 3070 cm$^{-1}$ but that of **B** has absorptions at 3040 and 2940 cm$^{-1}$. Spectrum **A** is assigned to **I**, and **B** to **II**.

➤
**See Table 9.6 in H&C**

11.  (a) The bands around 3000 cm$^{-1}$ may be assigned to C–H stretches involving both $sp^2$ and $sp^3$ C atoms; the strong absorptions at 1745 and 1715 cm$^{-1}$ can be assigned to stretches of the two different C=O groups; the remaining absorptions are in the fingerprint region and cannot readily be assigned at our level of interpretation. (b) Aspirin and cocaine have diagnostic IR spectra; the compounds are structurally dissimilar and their spectra cannot be confused; all that the customs officers need is an IR spectrometer!

➤
***NB*: You should *not* assume that two different C=O groups in a molecule will necessarily give rise to two IR spectroscopic absorptions**

---

**Problem set 9.4**

1.  (a) 30 nm; (b) $3.5 \times 10^{-2}$ nm; (c) 440.5 nm
2.  (a) and (b); look at Figure 9.20 in H&C
3.  Refer to Appendix 4 in H&C; vacuum-UV is at higher energy ($\Delta E$ large) than the visible region of the electromagnetic spectrum.
4.  (a) Highest occupied molecular orbital; lowest unoccupied molecular orbital; (b) non-bonding molecular orbital.
5.  See the first sub-section of Section 9.9 in H&C.
6.  Refer to Figure 9.20 in Section 9.9 in H&C; I$^-$, near-UV; [C$_2$O$_4$]$^{2-}$, near-UV; [N$_3$]$^-$, near-UV; H$_2$O, vacuum- (or far) UV; HC≡CH, vacuum-UV; [Ni(H$_2$O)$_6$]$^{2+}$, visible; C$_6$H$_5$N=NC$_6$H$_5$, near-UV.
7.  You might find it helpful to plot the absorbance data to see the cut-off points of the solvents. (a) The tail of the absorption due to cyclohexane may mask the absorption of benzene at 183 nm but the bands at 204 and 256 nm should be visible, the former as a shoulder on the side of side of the absorption band of cyclohexane. *However*, you have not been given information about the relative extinction coefficients of benzene and cyclohexane and you need these data to be determine whether the absorption at 183 nm will be masked or not. (b) DMF and acetone are polar; simple ionic salts are not soluble in non-polar solvents and polar liquids tend to be miscible with other polar liquids, likewise non-polar with non-polar.

➤
**Intermolecular interactions:see Sections 1.21 and 11.8 in H&C**

(c) Cyclohexane has the lower cut-off point and all the absorption bands of pyrazine should be visible. (d) Any one of the solvents is suitable.

---

**Problem set 9.5**

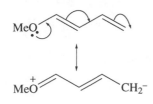

**9.5.1** Resonance structures for MeOCH=CHCH=CH$_2$

1. A molecule having three or more C=C double bonds.
2. Delocalized in (a), (d) — there must be alternating C=C and C–C bonds.
3. (a) Longer wavelength corresponds to a smaller energy change for the electronic transition ($E = h\nu$; $c = \nu\lambda$); as the number of C=C bonds increases, the degree of conjugation increases and the HOMO–LUMO gap (i.e. the energy of the electronic transition) becomes smaller; (b) for $n = 3$ or 4, near-UV; for $n = 5$ or 6, visible.
4. Absorbs the complementary colour, blue; see Table 9.2 in H&C. For an absorption at $\lambda = 469$ nm, the *transmitted light* is orange.
5. (a) A shift to longer wavelength; (b) lone pair donation (Figure 9.5.1) increases conjugation and lowers the HOMO-LUMO energy difference.

---

**Problem 9.6**

➤ **Beer-Lambert Law: equation 9.1 on page 75 of the workbook**

1. Absorbance increases during the reaction corresponding to one of the products absorbing at 470 nm; this is in the visible region and chloride ions are colourless; the organic product must absorb at 470 nm; the data in Table 9.2 in H&C indicate that the product appears red.
2. Use the Beer-Lambert Law to determine the concentration of [I$_3$]$^-$ ions; this equals the concentration of I$_2$ formed *provided that excess* I$^-$ is present and all the I$_2$ is in the form of [I$_3$]$^-$.
3. Figure 9.6.1 plots the data given in the question; extrapolation of the lines gives a point of maximum absorbance which corresponds to a solution composition of 5 cm$^3$ 0.002 M Cr(NO$_3$)$_3$ and 5 cm$^3$ 0.002 M H$_4$L; thus, in the complex, moles of Cr(III) = moles L$^{4-}$, and the formula is [CrL]$^-$.

**9.6.1** Plot of the variation of absorbance as a function of solution composition to find the stoichiometry of the complex formed between Cr(III) ions and L$^{4-}$. The lines are extrapolated to give the maximum absorbance and this gives the ratio of moles of Cr(III) : L$^{4-}$.

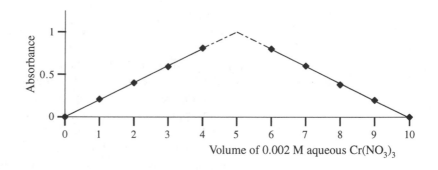

---

**Problem set 9.7**

1. Possible products are CH$_3$CH$_2$CH$_2$I (**A**) or CH$_3$CHICH$_3$ (**B**); only **A** has three different carbon environments.
2. (a) See Figure 9.7.1; (b) isomer **I** is structure **C**; isomer **II** could be structure **A** or **B** so assignment is ambiguous; isomer **III** is structure **D**.
3. Reaction between H$_2$C=CH$_2$ and Cl$_2$ gives CH$_2$ClCH$_2$Cl with one carbon environment; addition of Cl$_2$ to HC≡CH may lead to either the (*E*)- or (*Z*)-isomers of HClC=CHCl, each with one *sp$^2$* carbon environment.

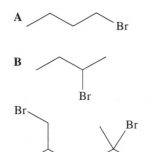

**A** ─ Br

**B** ─ Br

Br ─ **C**   Br ─ **D**

**9.7.1** Four structural isomers of $C_4H_9Br$.

$$\begin{matrix} H & & Me \\ & C=C & \\ Br & & H \end{matrix} \quad \begin{matrix} H & & H \\ & C=C & \\ Br & & Me \end{matrix}$$

(E)-isomer        (Z)-isomer

**9.7.2** Geometrical isomers of HBrC=CHMe.

4.  (a) $CH_2$ groups, $sp^3$ carbon; $C{\equiv}N$, $sp$ carbon; (b) only the signal at $\delta$ 119 can be unambiguously assigned (to the equivalent $sp$ carbon centres).

5.  Addition of $H_2O$ to $CH_3C(Me){=}CH_2$ could give $CH_3CH(Me)CH_2OH$ or $CH_3C(Me)(OH)CH_3$, the latter being Markovnikov addition; the $^{13}C$ NMR spectrum of $CH_3CH(Me)CH_2OH$ would show 4 signals (1:1:1:1), while that of $CH_3C(Me)(OH)CH_3$ is predicted to have 2 signals (3:1). Experimental data therefore confirm Markovnikov addition.

6.  (a) See Figure 9.7.2; (b) the identity of the product (i.e. which geometrical isomer is present) could be determined if $^{13}C$ NMR spectra of authentic samples were available for comparison.

7.  1,1,2,2-tetrachloroethane; (1,1,1,3-tetrachloroethane has 2 carbon environments).

8.  $\delta$ 170.3 is assigned to the C=O carbon centres (equivalent), $\delta$ 8.4 to the $CH_3$ groups (equivalent), and $\delta$ 28.7 to the two $CH_2$ groups.

9.  Signal at $\delta$ 162 is assigned to the C=O carbon; the signals at $\delta$ 30 and 36 are due to two *different* methyl carbon centres and this means that there is no free rotation about the C–N bond; you can understand why this is by considering resonance structures for DMF. [*Hint*: Can you invoke C=N double bond character?]

---

**Problem set 9.8**

1.  (a) Assignments: $\delta$ 11.4, OH; $\delta$ 2.1, $CH_3$; (b) relative integrals of the peaks, *but care is needed* — signals due to the OH of carboxylic acids are broadened and the integrals may not be consistent with the expected 1:3.

2.  Assignments: $\delta$ 2.2, OH; $\delta$ 1.5, Me; $\delta$ 2.4, ${\equiv}C{-}H$.

3.  $\overset{b}{\phantom{|}}\ \overset{a}{\phantom{|}}\ \overset{b}{\phantom{|}}$
    $CH_3CHICH_3$
    There are two H environments since the two $CH_3$ groups are equivalent; proton *a* couples to 6 equivalent *b* to give a septet; protons *b* appear as a doublet (coupling to one *a*).

4.  $\overset{a}{\phantom{|}}\ \overset{b}{\phantom{|}}$
    $CHCl_2CH_2Cl$
    Proton *a* couples to two *b* to give a triplet; coupling of protons *b* to one *a* gives a doublet.

5.  $CH_3CH_2Br$; the terminal $CH_3$ protons come at higher field ($\delta$ 1.7) than the $CH_2$ group ($\delta$ 3.4); signal at $\delta$ 1.7 appears as a triplet (coupling to 2 equivalent protons), and signal at $\delta$ 3.4 appears as a quartet (coupling to 3 equivalent protons).

6.  Possible isomers are $CCl_3CH_2CH_3$ (**A**), $CHCl_2CHClCH_3$ (**B**), $CH_2ClCH_2CHCl_2$ (**C**), $CH_3CCl_2CH_2Cl$ (**D**), $CH_2ClCHClCH_2Cl$ (**E**); both **A** and **D** have 2 proton environments in a ratio 3 : 2 but there would be coupling in **A** to give a triplet and a quartet; the protons are further apart in **D** and singlets are expected; the isomer present is $CH_3CCl_2CH_2Cl$.

7.  Assignments: $\delta$ 1.2, $CH_3$ (doublet, rel. integral 6); $\delta$ 1.6, OH (broad singlet rel. integral 1 but may not be accurately measured); $\delta$ 4.0, CH (septet, rel. integral 1).

8.    *a  b  c*
      $CH_3CH_2CH_2NO_2$
      (a) Protons *a* couple to 2 equivalent *b* to give a triplet (at $\delta$ 1.0); protons *c* couple to 2 equivalent *b* to give a triplet (at $\delta$ 4.4); protons *b* couple to 2 protons *c* and 3 protons *a*, but the observed sextet shows that the coupling constants are equal; (b) yes, $CH_3CH(NO_2)CH_3$ contains two proton environments in a ratio 1 : 6 (observe a doublet and a septet).

---

**Problem set 9.9**

1.    (a) $\delta$ 12.2; (b) and (c) see Figure 9.9.1.
2.    $CH_3CH_2CH_2CH_2Cl$: 4 signals (3:2:2:2); $CH_3CH_2CHClCH_3$: 4 signals (3:2:1:3); $CH_3CCl(Me)CH_3$: 1 signal (singlet); relative integrals should be sufficient, but coupling patterns will confirm assignments.
3.    (a) Trigonal bipyramid; 2 (axial and equatorial); (b) $PF_5$ undergoes Berry pseudo-rotation (see Section 5.12 in H&C) giving five apparently equivalent F atoms; $^{19}F$ nuclei couple with $^{31}P$ nucleus to give a sextet in the $^{31}P$ NMR spectrum; (c) $^{19}F$ spectrum has one signal, a doublet due to $^{19}F-^{31}P$ coupling.
4.    (a) Octahedral; (b) there are 6 equivalent F atoms; $^{19}F-^{31}P$ coupling gives a doublet in the $^{19}F$ NMR spectrum, and a septet in the $^{19}F$ NMR spectrum.
5.    In the $^1H$ NMR spectrum, the $CH_2$ protons appear as a quartet due to $^1H-^{19}F$ coupling; the broad peak is due to the OH proton.
6.    The 3 equivalent $^{19}F$ nuclei must couple to *both* types of $^{13}C$ to give two quartets, i.e. *long range coupling* is observed; the quartet with the *larger* value of $J_{CF}$ is assigned to the carbon centre *directly attached* to the fluorine atoms (i.e. the $CF_3$ carbon).

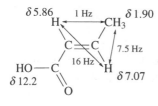

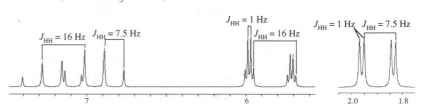

**9.9.1**    $^1H$ NMR spectroscopic chemical shift values and $^1H-^1H$ coupling in (*E*)-but-2-enoic acid.

---

**Problem set 10.1**

1.    Half-lives are constant, 11.5 s; reaction is first order with respect to alcohol.
2.    First order with respect to $[MnO_4]^-$; plot of ln (abs) against time is linear.
3.    Linear; the 'characterisic half-life' means the half-life is constant and so decay is first order; the decay of a daughter nuclide may complicate matters.

➤
**Radioactive decay:
see Box 10.2 in H&C**

4.    (a) Pseudo-first order rate constant; (b) $x = 1$; $k = 169$ $dm^3$ $mol^{-1}$ $min^{-1}$; determined from a plot of log $k_{obs}$ against log $[A]$, or you can show that $x = 1$ from a plot (linear) of $k_{obs}$ against $[A]$, with a gradient $k = 174$ $dm^3$ $mol^{-1}$ $min^{-1}$; (c) second order overall.
5.    (a) Oxalate ion; (b) the concentration of a component that is in vast excess effectively remains constant during the reaction and the rate is seen to depend only upon the component, the concentration of which varies during the reaction — in this case $[MnO_4]^-$; (c) first order; a plot of ln $[MnO_4^-]$ against time is linear.

6.  At the *end* of the reaction ($t = \infty$, from extrapolation in Figure 10.6, page 93), $[Y]_\infty = \frac{1}{2}[X]_0$; i.e. $2X \rightarrow Y$.

7.  (a) A plot of $\frac{1}{[A]}$ against time is linear, showing that the reaction is second order with respect to A; (b) Rate $= k[A]^2[B]^0 = k[A]^2$; (c) $k = 0.333$ dm$^3$ mol$^{-1}$ min$^{-1}$ or $5.55 \times 10^{-3}$ dm$^3$ mol$^{-1}$ s$^{-1}$.

8.  Firstly, note that the orders with respect to **I** and **II** are the same; (a) plot is non-linear showing $n \neq 0$;

    (b) a plot of $\frac{1}{[II]}$ against time is linear showing that the reaction is second order overall; *however* these data illustrate that it may be difficult to decide if a plot is linear — you may be able to decide by looking at the correlation coefficient;

    for the plot of $\frac{1}{[II]}$ against time, the correlation coefficient is 0.9927 (i.e. $\approx 1$) indicating that the data fit a linear relationship, while for the ln [**II**] versus time plot, the coefficient is 0.9655; in practice, data for $t > 6000$ s are required to minimize ambiguity; (c) $k = 9.6 \times 10^{-4}$ dm$^3$ mol$^{-1}$ s$^{-1}$.

9.  (a) From a plot of log (initial rate) against log $[NO_3^-]_0$, the gradient $= n = 1$; (b) the *kinetics* of the reaction indicate that the nitrate ion is involved in a rate determining step; note that the *equation* for the reaction between $H_2$ and $[MnO_4]^-$ gives no information about the involvement of the catalyst.

10. (a) From a tangent drawn to the curve, initial rate $\approx 1 \times 10^{-6}$ mol dm$^{-3}$ s$^{-1}$; (b) from a plot of log (rate) against log $[A]_0$, gradient $= 1.85$, giving $n = 2$ (the nearest integer); (c) pseudo-second order rate constant, $k_{obs}$, is found from intercept on the log (rate) axis: log $k_{obs} = -4.14$, $k_{obs} = 7.24 \times 10^{-5}$ dm$^3$ mol$^{-1}$ s$^{-1}$; this is not the same as $k$ in the rate equation (see question 4).

11. (a) $A_{corr} = A - 0.108$; (b) the plot is a curve, showing that the reaction is *not* zero order with respect to isomer **I**; (c) first order; a plot of ln $A_{corr}$ against time is linear; (d) from the gradient of the graph in (c), $k = 0.195$ day$^{-1}$, or, in SI units, $2.26 \times 10^{-6}$ s$^{-1}$.

12. (a) The plot is a curve (Figure 10.1.1) although it would be very easy to fit the data to a straight line; correlation coefficient $= 0.99$!; the decay should be followed for a longer period to clarify ambiguity; (b) plot of ln $A$ against time is linear; (c) $k = 2.54 \times 10^{-3}$ s$^{-1}$ from the gradient of the plot in part (b).

**10.1.1** For question 12(a).

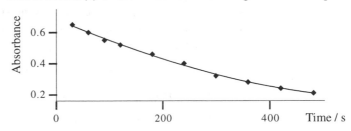

**Problem set 10.2**

1.  Two values do not allow for experimental error. A greater number of readings permit $E_{activation}$ to be found from a plot of ln $k$ against $\frac{1}{T}$ (see question 2).

2.  $E_{activation} = $ (a) 147.0; (b) 120.6; (c) 134.9 kJ mol$^{-1}$; an average of the three readings would give 134.2 kJ mol$^{-1}$ (compared to 132.7 kJ mol$^{-1}$ obtained

in worked example 10.6 in H&C using all the data) which is a better estimate than using only two points of experimental data.

3.  $E_{activation} = 82.6 \text{ kJ mol}^{-1}$

4.  (a) The product is an ionic species (the reactants are covalent species); the increase in concentration of product will lead to an increase in the conductance of the solution; (b) $E_{activation} = 56.7 \text{ kJ mol}^{-1}$; (c) the data given fit a straight line equation:
$$\ln k = 19.08 - \frac{6821}{T}$$
from which $k(298 \text{ K}) = 2.22 \times 10^{-2} \text{ dm}^3 \text{ mol}^{-1} \text{ min}^{-1}$.

---

**Problem set 10.3**

1.  (a) Unimolecular; (b) bimolecular; (c) bimolecular; (d) unimolecular.

2.  Assuming that $k$ represents each rate constant (but not implying that the values are the same!): (a) Rate = $k[H_2]$; (b) Rate = $k[H^\bullet][Br_2]$; (c) Rate = $[Cl^\bullet]^2$; (d) Rate = $k[Me_3CBr]$.

3.  (a) Rate = $k[ClO^\bullet]^2$; Rate = $k[Cl_2O_2]$; Rate = $k[Cl^\bullet][O_3]$; (b) the first and third reactions are bimolecular.

4.  (a) In an elementary step, the stoichiometry gives the orders directly; (b) no relationship can be assumed.

5.  (a) $\dfrac{d[D]}{dt} = k_1[A]$ ; (b) $\dfrac{d[D]}{dt} = k_1[A] - k_2[D]$ ;

    (c) The steady-state approximation assumes that $\dfrac{d[D]}{dt} = 0$.

    So, from part (b): $k_1[A] - k_2[D] = 0 \qquad [D] = \dfrac{k_1[A]}{k_2}$

    The rate of formation of B is: $\dfrac{d[B]}{dt} = k_2[D] = k_2\left(\dfrac{k_1[A]}{k_2}\right) = k_1[A]$

---

**Problem set 11.1**

1.  (a) $CH_3CO_2H(aq) + H_2O(l) \rightleftharpoons [H_3O]^+(aq) + [CH_3CO_2]^-(aq)$
    (b) $HNO_3(aq) + H_2O(l) \rightarrow [H_3O]^+(aq) + [NO_3]^-(aq)$  (strong acid)
    (c) $H_2SO_3(aq) + H_2O(l) \rightleftharpoons [H_3O]^+(aq) + [HSO_3]^-(aq)$
    $[HSO_3]^-(aq) + H_2O(l) \rightleftharpoons [H_3O]^+(aq) + [SO_3]^{2-}(aq)$

2.  3.3

3.  $1.78 \times 10^{-4} \text{ mol dm}^{-3}$

4.  See Figure 11.1.1; you cannot assign particular $H^+$ dissociations.

5.  Chloroacetic acid; smaller $pK_a$ means more dissociation occurs in solution.

6.  Hypochlorous acid; larger $K_a$ means more dissociation occurs in solution.

7.  $5.0 \times 10^{-3}$ moles

8.  $1.8 \times 10^{-3} \text{ mol dm}^{-3}$

9.  $4.5 \times 10^{-6} \text{ mol dm}^{-3}$ ($K_a = 4.0 \times 10^{-10} \text{ mol dm}^{-3}$)

10. They are equal.

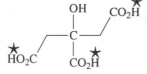

**11.1.1** Protons marked ★ in carboxylic acid groups ($-CO_2H$) dissociate.

---

**Problem set 11.2**

1.  (a) 0.012 moles; (b) 60 cm³

2.  $6.3 \times 10^{-6} \text{ mol dm}^{-3}$

3.  8.75

4.  (a) $NH_3(aq) + H_2O(l) \rightleftharpoons [NH_4]^+(aq) + [OH]^-(aq)$
    (i.e. it forms *ammonium hydroxide*); (b) $1.63 \times 10^{-3}$ mol dm$^{-3}$

5.  Make the data directly comparable: **A**: $pK_b = 9.10$; **B**: $pK_b = 10.15$; **C**: $pK_b = 8.88$. Smallest value corresponds to the greatest degree of dissociation; base strengths: **C > A > B**.

---

**Problem set 11.3**

1.  (a) 1.0; (b) yes; HCl(aq) is fully dissociated
2.  2 g NaOH = 0.05 moles; $[OH]^- = 0.05$ mol dm$^{-3}$; pH = 12.7
3.  $[H^+] = 9.2 \times 10^{-4}$ mol dm$^{-3}$; pH = 3.0
4.  pH changes from 1 to 2, a change of +1
5.  $[H^+] = 2 \times [H_2SO_4]$; pH = 0.30
6.  pH = 2.1; (remember, total volume of solution = 65 cm$^3$)
7.  (a) $Ba(OH)_2(aq) + 2HNO_3(aq) \rightarrow Ba(NO_3)_2(aq) + 2H_2O(l)$; (b) pH = 11.7
8.  pH = 12.4; even though the acid is weak, by Le Chatelier's principle, neutralization of some H$^+$ will cause more acid to dissociate; there is sufficient alkali present to neutralize all the acid.
9.  pH = 11.1
10. (a) $NH_3(aq) + HNO_3(aq) \rightarrow [NH_4]^+(aq) + [NO_3]^-(aq)$; (b) pH = 2.0

---

**Problem set 11.4**

1.  (a) Phenolphthalein, phenol red, bromocresol purple, bromocresol green; (methyl orange changes too early); (b) phenolphthalein; (probably phenol red); (c) in this titration we reach the higher pH value *first* making bromocresol purple, bromocresol green and methyl orange suitable.
2.  Two end-points require two indicators, unless a single indicator changes colour twice *but* these colour changes must correspond to the ranges pH 4–5 and 8.5–9 (see Figure 11.4, page 102)
3.  $[In^-] : [HIn] = $ (a) 34.7, (b) $1.00 \times 10^{-3}$
4.  Dibasic, $H_2In$

---

**Problem set 12.1**

1.  (a) $CO_2$ released to the surroundings; work done *by the system*; (b) negative.
2.  $w = -227$ J (per 0.1 mole $Na_2CO_3$).
3.  Work done by the system.
4.  $C_6H_{14}(l) + 9\frac{1}{2}O_2(g) \rightarrow 6CO_2(g) + 7H_2O(l)$; work is done *by the surroundings* since the number of moles of gas decreases.
5.  $2NaN_3(s) \rightarrow 2Na(s) + 3N_2(g)$; 3 moles of $N_2$ produced; $w = -7434$ J (per 2 moles of $NaN_3$).
6.  $C_2H_4(g) + H_2(g) \rightarrow C_2H_6(g)$; $w = +2993$ J (per 1.2 moles of ethene).

---

**Problem set 12.2**

1.  218.6 kJ per mole of H
2.  (a) 570.4 kJ per mole of HF; (b) 573.3 kJ per mole of HF.
3.  (a) $CO(g) + \frac{1}{2}O_2(g) \rightarrow CO_2(g)$; (b) $-283.15$ kJ per mole of CO
4.  The variation of $C_p$ over the range 298–500 K should be small for diatomics (a), (c) and (e), larger for the triatomic (d), and very significant for the polyatomics (b) and (f); [see Figure 12.6 in H&C].

---

**Problem set 12.3**

1. (a) –243 kJ per mole of $C_2H_2$; (b) –33 kJ per mole of $C_2H_4$; (c) –514 kJ per mole of reaction; (d) +6 kJ per mole of $C_2H_5OH$; (e) +131 kJ per mole $CaCO_3$; (f) –402 kJ per mole of Mg.
2. (a), (b), (c), (f)
3. (a) 0 kJ $mol^{-1}$; (b) 0 kJ $mol^{-1}$; (c) 0 kJ $mol^{-1}$
4. Probably not.
5. (a) NiO, $SnO_2$, FeO; (b) no; (c) no. [*Hint*: See worked example 12.4 in H&C.]
6. (a) $NO_2$, $SO_2$, $GeCl_4$; (b) CO, $SO_2$, $GeCl_4$; (all negative $\Delta_f G°$); (c) $SO_2$; (d) $GeCl_4$; (e) at 300 K, the formation of $SO_2$ from $S_8(s)$ and $O_2(g)$ is significantly more favoured than that of CO from C(gr) and $O_2(g)$, but this difference becomes smaller as the temperature is raised and at 1500 K, the values of $\Delta_f G°$ are almost equal; (f) no.

**Problem set 12.4**

1. (a) $3.75 \times 10^{-16}$ (no units); (b) equilibrium lies well to the left-hand side.
2. 298 K: equilibrium lies to the left-hand side; +1.7 kJ per mole of HI; 800 K: equilibrium lies to the right-hand side; –12.3 kJ per mole of HI.
3. Plot $\Delta_f G°(T$ K) against ln $K$ [another point is $\Delta_f G°(T$ K) = 0 = ln $K$]; the gradient of the line = $-T \times R$, giving $T$ = 298 K.
4. (a) –117 kJ per mole $H_2O_2$; (b) $3.22 \times 10^{20}$ (dimensionless — determined from a ln $K$ value); (c) kinetically stable, see Chapter 10 of H&C.
5. –301 kJ per mole of $SO_2$.
6. (a) $2.29 \times 10^9$ (dimensionless — determined from a ln $K$ value); (b) yes.

**Problem set 12.5**

1. 192 J $K^{-1}$ $mol^{-1}$
2. $S°(H_2O$, g, 373 K)
3. 25.5 J $K^{-1}$ $mol^{-1}$
4. 85.3 J $K^{-1}$ $mol^{-1}$
5. (a) 109 J $K^{-1}$ $mol^{-1}$; (b) value obtained for $CCl_4$ is consistent with Trouton's rule but the higher (anomalous) value for $H_2O$ reflects the significant hydrogen bonding between $H_2O$ molecules in the liquid state.
6. Ge is a crystalline solid at 298 K; the discontinuity in the curve corresponds to the melting point (1211 K); $S°$(liquid) > $S°$(solid).
7. Units of $\Delta S°$(298 K) are J $K^{-1}$ per mole of reaction: (a) –439; (b) –152; (c) +164; (d) +129; (e) +20.
8. Favoured by entropy: (c), (d), (e); relate the values to the number of moles of gas produced or consumed; gases have larger $S°$ values than liquids or solids; it is harder to assess reactions such as (e).

**Problem set 12.6**

1. (a) Mn(s) + $Cl_2(g)$ → $MnCl_2(s)$; (b) –136 J $K^{-1}$ $mol^{-1}$; (c) –137 J $K^{-1}$ $mol^{-1}$
2. (a) 14 J $K^{-1}$ per mole of reaction, or 7 J $K^{-1}$ $mol^{-1}$ per mole of HF; (b) yes, but the balance between $\Sigma S°_{products}$ and $\Sigma S°_{reactants}$ is small; (c) no, because the $T\Delta S°$ term is very small; experimental values are listed in Appendix 11.
3. (a) –186.3 J $K^{-1}$ per mole of $CH_3NH_2$; +33.0 kJ $mol^{-1}$
4. (a) 2C(graphite) + $H_2(g)$ + $Cl_2(g)$ → $CH_2=CCl_2(l)$; (b) –163.9 J $K^{-1}$ $mol^{-1}$; (c) no; (d) $\Delta_f H°$(298 K) = –24 kJ $mol^{-1}$; $\Delta_f G°$(298 K) = +24 kJ $mol^{-1}$; $\Delta_f G°$

shows that the formation is not favourable although $\Delta_f H°$ reveals an exothermic reaction. (e) The large negative value of $\Delta S$ is the critical in making $\Delta_f G°$ positive even though $\Delta_f H°$ is negative.

5.  (a) From $\Delta_f H°$, formation of $NH_3$ appears to be favoured at both temperatures; (b) from $\Delta_f G°$, $NH_3$ formation is favoured at 298 K but not at 500 K; (c) $\Delta_f S°(NH_3, g)$: 298 K, $-99.0$ J $K^{-1}$ $mol^{-1}$, 500 K, $-109.4$ J $K^{-1}$ $mol^{-1}$; values are quite similar; $T\Delta S$ terms: 298 K, $-29.5$ kJ $mol^{-1}$, 500 K, $-54.7$ kJ $mol^{-1}$; values differ enough to swing the balance between the $\Delta H$ and $T\Delta S$ terms making the reaction favourable at 298 K but not at 500 K.

---

**Problem set 12.7**

1.  (a) $Fe(s) + 2H^+(aq) \rightarrow Fe^{2+}(aq) + H_2(g)$
    (b) $2Ag^+(aq) + Zn(s) \rightarrow 2Ag(s) + Zn^{2+}(aq)$; a good way to grow silver crystals!
    (c) $2Ce^{4+}(aq) + 2I^-(aq) \rightarrow 2Ce^{3+}(aq) + I_2(aq)$
2.  (a) 0.44 V; (b) 1.56 V; (c) 1.18 V
3.  (a) Oxidation: $V^{2+}(aq) \rightarrow V^{3+}(aq) + e^-$; reduction: $2H^+(aq) + 2e^- \rightarrow H_2(g)$;
    (b) $2V^{2+}(aq) + 2H^+(aq) \rightarrow 2V^{3+}(aq) + H_2(g)$; (c) $-0.26$ V
4.  (a) $Pb^{2+}(aq) + 2e^-(aq) \rightarrow Pb(s)$;  (b) $Zn(s) \left| Zn^{2+}(aq) \vdots Pb^{2+}(aq) \right| Pb(s)$
    (c) 0.63 V

---

**Problem set 12.8**

1.  (a) $Cl_2(aq) + Zn(s) \rightarrow 2Cl^-(aq) + Zn^{2+}(aq)$
    (b) $2H^+(aq) + Mg(s) \rightarrow H_2(g) + Mg^{2+}(aq)$
    (c) $Cl_2(aq) + 2Br^-(aq) \rightarrow 2Cl^-(aq) + Br_2(aq)$
    (d) $H_2O_2(aq) + 2H^+(aq) + 2I^-(aq) \rightarrow 2H_2O(l) + I_2(aq)$
    (e) $[IO_3]^-(aq) + 6H^+(aq) + 6I^-(aq) \rightarrow I^-(aq) + 3H_2O(l) + 3I_2(aq)$
    or  $[IO_3]^-(aq) + 6H^+(aq) + 5I^-(aq) \rightarrow 3H_2O(l) + 3I_2(aq)$
2.  $E°_{cell}$ (V): (a) 2.12; (b) 2.37; (c) 0.27; (d) 1.24; (e) 0.66; $\Delta G°$(298 K) (kJ $mol^{-1}$): (a) $-409$; (b) $-457$; (c) $-52$; (d) $-239$; (e) $-382$
3.  (a) $Cl_2$; (b) $H^+$; (c) $Cl_2$; (d) $H_2O_2$; (e) $[IO_3]^-$
4.  (a) Oxidizing agent;
    (b) $H_2O_2(aq) + 2H^+(aq) + 2Fe^{2+}(aq) \rightarrow 2H_2O(l) + 2Fe^{3+}(aq)$;
    (c) Reducing agent;
    (d) $5H_2O_2(aq) + 2[MnO_4]^- + \cancel{6}H^+(aq) \rightarrow 2Mn^{2+}(aq) + 8H_2O(l) + 5O_2(g) + \cancel{10H^+(aq)}$

➤
**Cancel 10H$^+$ on the left-
and right-hand sides of
the equation**

---

**Problem set 12.9**

1.  0.30 V
2.  $-0.06$ V (from 0.80 to 0.74 V)
3.  0.21 mol $dm^{-3}$
4.  $-0.01$ V (from $-0.36$ to $-0.37$ V)
5.  $+0.68$ V

---

**Problem set 12.10**

1.  (a) $[Ba^{2+}][SO_4^{2-}]$; (b) $[Mg^{2+}]^3[PO_4^{3-}]^2$; (c) $[Fe^{3+}][OH^-]^3$ ; (d) $[Pb^{2+}][I^-]^2$
2.  $3.58 \times 10^{-3}$ mol $dm^{-3}$
3.  $1.1 \times 10^{-5}$ mol $dm^{-3}$
4.  $8.5 \times 10^{-17}$ $mol^2$ $dm^{-6}$
5.  $5.3 \times 10^{-13}$ $mol^2$ $dm^{-6}$

## Problem set 13.2

**13.2.1** There are three fluorine environments in $[Sb_2F_{11}]^-$.

1. $[AsF_6]^-$ is octahedral; all $^{19}F$ nuclei are equivalent, giving a singlet in the $^{19}F$ NMR spectrum.

2. $[PF_6]^-$ is octahedral; all $^{19}F$ nuclei are equivalent and couple to $^{31}P$ giving a septet.

3. See Figure 13.2.1; three environments, labelled $a$ to $c$ in the figure.

4. $PF_5$ is fluxional on the NMR spectroscopic timescale at 298 K; $^{19}F$ appear equivalent, giving one signal which appears as a doublet due to coupling to $^{31}P$ nucleus.

## Problem set 14.1

1. (a) e.g. Reaction of butan-2-ol with $SOCl_2$, $PCl_3$ or $PCl_5$, or reaction of but-2-ene with HCl; reaction of but-1-ene with HCl would also give 2-chlorobutane as the favoured product; (b) addition of $Br_2$ to propene; (c) photolysis of $CH_4$ with $Cl_2$ but reaction would be non-specific also forming e.g. $CH_2Cl_2$, $CHCl_3$ and $C_2H_6$.

2. (a) $CH_3CH_2Li + LiCl$; (b) $LiCl + CH_3CH_2CH_2CH_2CH_2CH_3$ (coupling), or $CH_3CH_2Li + CH_3CH_2CH_2CH_2Cl$ (exchange); (c) $CH_3CH_2CH_2MgCl$; (d) $NaBr + CH_3CH_2OCH_2CH_3$.

3-4. For the mechanisms, see H&C, Section 14.5.

5. See Figures 14.7 and 14.8 in H&C, plus accompanying discussion.

6. $S_N1$ mechanism (with no complications).

7. See main text.

8. (a) $C_2H_5OH$ and KOH; (b) strong; (c) $CH_3CH=CH_2$ (other products might be $CH_3CH_2CH_2OH$ and, from the $[EtO]^-$ ion present, $CH_3CH_2OCH_2CH_2CH_3$.

9. (a) Mechanism is an $S_N2$ displacement; (b) rate of racemization = 2 × rate of exchange.

10. (a) Rate = $k[R_3CX]$; (b) no (see main text).

## Problem set 14.2

**14.2.1** The isomer is 2-methylpropan-1-ol.

**14.2.2** Structure of 2,5-dioxahexane.

1. See Figure 14.2.1.

2. (a) $CH_3(CH_2)_{14}CH_2OH$; (b) broad absorption ≈3300 cm$^{-1}$ due to O–H stretch; bands between 2800 and 3000 cm$^{-1}$ due to C–H stretches; lower frequency bands are in the fingerprint region.

3. (a) C–H stretching mode; (b) band at ≈1100 cm$^{-1}$; (c) an O–H stretch; the sample is wet.

4. See H&C, Section 14.7.

5. (a) Hydrogen bonds cannot form between $Et_2O$ molecules; intermolecular hydrogen bonding in butan-1-ol results in association in the liquid state and an increase in boiling point relative to the ether; (b) serious fire hazard and a risk of explosion due to the formation of peroxides.

6. See Figure 14.2.2; using labels from figure: $\delta\,58.6 = a$; $\delta\,72.3 = b$.

7. $\overset{a}{CH_3}\overset{b}{SCH_2}\overset{c}{CH_3}$   $\delta\,2.1 = a$, $\delta\,2.5 = b$, $\delta\,1.3 = c$.

8. (a) $CH_3CH=CHCH_3$, elimination of $H_2O$; (b) $CH_3CH_2CH_2Cl + SO_2 + HCl$; (c) $Na[Me_3CO] + \frac{1}{2}H_2$; (d) $[H_3NCH_2CH(NH_3)CH_3]Br_2$; (e) $CH_3CH_2CH_2NH_2$; (f) $Me_2C=O$.

9. See Sections 8.14, 8.18 and 14.9 in H&C.

10. IR spectrum of $EtNH_2$ would show a broad absorption for the N–H stretch in the range 3000-3500 $cm^{-1}$, but $Et_3N$ would not.

11. –372 kJ per mole of $(C_4H_9)_2O$

12. (a) $CH_3CH_2OH + CH_3CH_2Br$; (b) $CH_3CMe_2CMe_2Cl$; (c) no reaction; (d) $CH_3CHO$; (e) $HO_2CCO_2H$; (f) $[Et_4N][CH_3CO_2] + H_2O$

---

**Problem set 15.1**

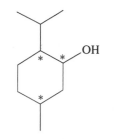

**15.1.1** Asymmetric carbon centres in menthol.

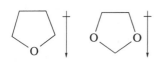

**15.1.2** Structures of and directions of molecular dipoles in THF and 1,3-dioxolane.

1. 3-Membered ring is too small to permit any degree of C–C bond rotation; in *cyclo*-$C_{10}H_{20}$, C–C bond rotation is restricted but sufficient to allow ring flexibility.

2. See Figure 15.4 in H&C.

3.

(a)    (b)    (c)    (d)    (e)

4. (a) 3; see Figure 15.1.1; (b) O–H stretch 3300 $cm^{-1}$; C–H stretches 3000–2800 $cm^{-1}$; absorptions below 1500 $cm^{-1}$ are in the fingerprint region; (c) each would become axial.

5. $\delta$ 211 is C=O carbon, position C(1); $\delta$ 25 (relative integral 1) assigned to unique $CH_2$ group, position C(4).

6. (a) Tetrahydrofuran; (b) and (c) see Figure 15.1.2; net dipole moment in 1,3-dioxolane depends on the presence of both O atoms, see pages 121-122 of workbook. [*Comment*: 1,3-dioxolane is susceptible to acid hydrolysis.]

---

**Problem set 15.2**

1. (a) Methylcyclopropane; (b) $CH_3CH_2CH_3$ (ring opening); (c) chloro-cyclohexane + polychlorinated derivatives; (d) cyclohexane, with some cyclohexene; (e) 1-bromo-1-methylcyclohexane with 1-bromo-2-methylcyclohexane as the minor product.

2. Strained ring, addition reactions; see Section 15.2 in H&C.

3–4. See Section 15.4 in H&C.

5. (a) **A** = cyclohexanol; **B** = cyclohexanone; alcohol is oxidized to a ketone; (b) ≈1700 $cm^{-1}$ diagnostic of C=O group; 2800–3000 $cm^{-1}$ due to C–H stretches; (c) $C_6H_{11}OH(l) + 8\frac{1}{2}O_2(g) \rightarrow 6CO_2(g) + 6H_2O(l)$

---

**Problem set 15.3**

1. Delocalized bonding: see Section 15.5 in H&C.

2. (a) and (b) See Figure 15.17 in H&C and accompanying discussion.

3. Molecular formula of compound with $M_r$ 92 = $C_7H_8$, and of that with $M_r$ 108 = $C_7H_8O$; **A** = $C_6H_5Me$ (toluene); **B** = $C_6H_5CH_2OH$ (benzyl alcohol); **C** = $C_6H_5OMe$ (anisole); in the $^1H$ NMR spectra, the multiplets around $\delta$ 7.2 are due to protons attached directly to the ring; in **A** and **C**, singlet is due to Me group; in **B**, singlet at $\delta$ 4.5 is assigned to $CH_2$ group and the

**15.3.1**
The
structures
of 1,3,5-
and 1,2,4-
$Me_3C_6H_3$.

broad signal to the OH proton.

4.  See Figure 15.3.1; 1,3,5-isomer has 3 carbon environments (3 signals of equal intensity in $^{13}C$ NMR spectrum); 1,2,4-isomer has 9 carbon environments (9 signals of equal intensity in $^{13}C$ NMR spectrum).

5.  (a) $C_6H_4Cl_2$; (b) 1,2- and 1,4-isomers; (c) 1,3-isomer (right) has four carbon environments in a ratio 1 : 2 : 2 : 1.

6.  (b) and (c): planar rings, delocalized $\pi$-system, obey the Hückel ($4n + 2$) rule.

---

**Problem set 15.4**

1.  (a) Cyclohexa-1,3-diene (Birch reduction); (b) chlorobenzene; (c) toluene in the first step, then benzoic acid, *but* Me group is activating and *ortho / para*-directing and so the formation of xylenes and the corresponding acids will occur; (d) nitrobenzene.

2.  (a), (b) and (d): see Sections 15.6–15.11 in H&C; (c) as question 1, part (c), followed by treatment with NaOH(aq).

3.  Scheme is as equations 15.40 to 15.42 in H&C with Br replacing Cl, and $FeBr_3$ replacing $AlCl_3$; the initial interaction between $FeBr_3$ and $Br_2$ is as we drew for $AlCl_3$ with $Cl_2$; the Wheland intermediate in Figure 15.4.1.

4.  Reaction of $CH_3CH_2CH_2Cl$ with $AlCl_3$ gives a primary carbenium ion that rearranges to a secondary carbenium ion:

$$CH_3CH_2\overset{+}{C}H_2 \rightarrow CH_3\overset{+}{C}HCH_3$$

**15.4.1** Wheland
intermediate in the reaction
of benzene with $Br_2$ in the
presence of $FeBr_3$.

Electrophilic substitution with benzene then gives $C_6H_5CHMe_2$ in preference to $C_6H_5CH_2CH_2CH_3$.

5.  Methyl group is susceptible to free radical substitution, but H atoms attached directly to the ring undergo electrophilic substitution; refer to the appropriate discussion in Section 15.9 in H&C.

6.  This question addresses the difference between phenol and benzyl alcohol; the latter behaves as a typical primary alcohol. Reactions of phenol: see Section 15.10 in H&C; reactions of an alcohol: see Section 14.10 in H&C.

➤ 7.  (a) (i) $NaNO_2$, HCl(aq), 273 K; (ii) $H_2O$; (iii) KOH; (iv) HCl; (v) $C_6H_5NH_2$; (b) azo-compound; (c) –N=N– group; absorbs in the ultraviolet–visible region; some azo-compounds are coloured and are used as dyes.

**Chromophores and azo-
dyes: see Sections 9.10
and 9.11 in H&C**

8.

pyridine (aq) + $H_2O$(l) ⇌ (aq) + $[OH]^-$(aq)

9.    (a)    (b)    (c)    (d)    (e)

N+          N+          N+          N+          N
H  Br⁻      H  CH₃CO₂⁻   I⁻         H  BF₄⁻      ↓
                                                 BF₃

---

**Problem set 16.1**

16.1.1 Resonance structures for [Hdmg]⁻.

1–2.  See Table 16.2 in H&C
3.    (a) Square planar, *cis* and *trans*-isomers; (b) square planar; (c) tetrahedral; (d) octahedral, *C*-bound to metal; (e) octahedral, three *N,N*'-bound ligands, optical isomers; (f) octahedral, *cis* and *trans*-isomers; (g) octahedral, three *O,O*'-bound ligands, optical isomers; (h) octahedral, two *N,N*'-bound ligands, *cis* and *trans*-isomers with two enantiomers of the *cis*-isomer.
4.    Free rotation is only possible about the inter-ring C–C bonds of the tpy ligand and this flexibility is insufficient to allow the formation of a *fac*-isomer; in the *mer*-isomer, each ligand is planar.
5.    (a) $Cr^{3+}$; (b) $[CrCl_2(H_2O)_4]^+$ is octahedral; *trans*- or *cis*-arrangements of Cl ligands; (c) hydration and ionization isomerism (see Section 16.5 in H&C).
6.    (a) and (c), and the *cis*-isomer of (d).
7.    (a) $Co^{3+}$; (b) steric effects ; the two bulky phosphine ligands occupy sites remote from one another.
8.    (a) See Figure 16.1.1; (b) $\pi$-conjugation in the ligand makes the ONCCNO-backbone planar, and hydrogen bonding between the two O–H····O units forces the coordination geometry at nickel to be planar.

---

**Problem set 16.2**

16.2.1 The structure of the dinuclear complex ion $[(H_2O)_4Cr(\mu\text{-OH})_2Cr(H_2O)_4]^{4+}$.

1.    (a) $1.3 \times 10^{-4}$ mol dm⁻³ ; $1.0 \times 10^{-2}$ mol dm⁻³;
      (b) $[M(H_2O)_6]^{3+}(aq) + H_2O(l) \rightleftharpoons [M(H_2O)_5(OH)]^{2+}(aq) + [H_3O]^+(aq)$
      (M = Ti or Fe); (c) $[Fe(H_2O)_6]^{3+}$; (d) $M^{3+}$ ions are strongly polarizing; the O–H bond in the $H_2O$ ligand becomes more polar than in free water and dissociation of H⁺ occurs; (e) $K_a$ for $[Fe(H_2O)_6]^{2+} < K_a$ for $[Fe(H_2O)_6]^{3+}$, and therefore $pK_a$ for $[Fe(H_2O)_6]^{2+} > [Fe(H_2O)_6]^{3+}$.
2.    (a) Fe(III); (b) $[Fe(H_2O)_6]^{3+}$; (c) purple; (d) $[Cr(H_2O)_6]^{3+}$.
3.    $[Co(H_2O)_6]^{2+}$; octahedral complex ion.
4.    (a) $2[Cr(H_2O)_5(OH)]^{2+}(aq) \rightleftharpoons [(H_2O)_4Cr(\mu\text{-OH})_2Cr(H_2O)_4]^{4+}(aq) + 2H_2O(l)$;
      (b) see Figure 16.2.1; (c) no, remains $Cr^{3+}$.

---

**Problem set 16.3**

➤

**Hard and soft metals and donor atoms: see Section 16.8 in H&C**

1.    (a) $K_3[Cr(NCS)_6]$; linkage isomers are possible but Cr(III) is a hard metal centre and is expected to prefer *N*-coordination; (b) $[Ni(C_2O_4)_2(H_2O)_2]^{2+}(aq)$; *trans* or *cis* isomers, with enantiomers of the *cis*-isomer; (c) $K_2[Zn(CN)_4]$; (d) $[Fe(bpy)_3]^{2+}(aq)$; enantiomers; (e) $[Fe(H_2O)_6]^{3+}(aq) + [NO_3]^-(aq)$
2.    Chelate effect: see 'didentate and polydentate ligands' in Section 16.6 in H&C; (a) 3; (b) 5; (c) 4.

3.   **A** is [CoBr(NH$_3$)$_5$][SO$_4$]; *non-coordinated* sulfate ion reacts with BaCl$_2$ to give a white precipitate of BaSO$_4$; **B** is [Co(NH$_3$)$_5$(SO$_4$)]Br ; *non-coordinated* bromide ion reacts with AgNO$_3$ to give a precipitate of AgBr.

4.   (a) Co(II); Co(III); (b) oxidizing agent; (c) [Co(NH$_3$)$_5$Cl]$^{2+}$; [Co(NH$_3$)$_5$(H$_2$O)]$^{3+}$.

5.   (a) [M(H$_2$O)$_6$]$^{2+}$(aq) + L $\rightleftharpoons$ [ML(H$_2$O)$_3$]$^{2+}$(aq) + 3H$_2$O(l)    where M = Ni or Cu; (b) equilibrium involving copper(II) ions; (c) *mer-* and *fac-*isomers.

6.   (a) 1 × 10$^{24}$ (dimensionless as determined from a log value);
    (b)
$$\beta_6 = \frac{[\text{Fe(CN)}_6{}^{4-}]}{[\text{Fe(H}_2\text{O)}_6{}^{2+}][\text{CN}^-]^6}$$

7.   $K_2 = 3.16 \times 10^3$ (dimensionless since determined from a log value); log $K_1 >$ log $K_2$ and so equilibrium for step 1 lies further to the right-hand side than does that for step 2.

---

**Problem set 16.4**

(a)

(b)

**16.4.1** Orbital energy diagrams showing 3*d* electronic configurations in (a) [FeF$_6$]$^{3-}$ and (b) [MnF$_6$]$^{3-}$.

1.   See Section 16.11 in H&C.

2.   See  Section 16.11 in H&C; (a) weak field; (b) strong field; (c) relatively weak field; (d) mid-field.

3.   Br$^-$ < Cl$^-$ < [OH]$^-$ < H$_2$O < NH$_3$ < [CN]$^-$; i.e. part of spectrochemical series.

4.   (a) $t_{2g}{}^5 e_g{}^0$; (b) $t_{2g}{}^3 e_g{}^1$; (c) $t_{2g}{}^3 e_g{}^0$; (d) $t_{2g}{}^1 e_g{}^0$; (e) $t_{2g}{}^2 e_g{}^0$; an electron cannot be promoted to an $e_g$ orbital if it leaves the orbital in the $t_{2g}$ set empty, hence no distinction between high- and low-spin states.

5.   (a) 4.90; (b) 1.73; (c) 2.83 BM.

6.   Diamagnetic; Sc(III) has no *d* electrons; it possesses a noble gas configuration.

7.   $n = 0$; possible values of $n$ are 0 to 6; other than 0, only $n = 6$ for a low spin complex gives a diamagnetic species but such a high charge is unrealistic.

8.   M = V (2.87 BM) or Ti (1.74 BM).

9.   (a) See Figure 16.4.1a; (b) 5.92 BM; (c) high-spin; (d) See Figure 16.4.1b; (e) F$^-$ is a relatively weak field ligand and this is consistent with high-spin complexes; however, you must *not assume* that all weak field ligands give rise to high-spin complexes.

---

**Problem set 16.5**

**16.5.1** Structure of PMe$_3$.

1.   See Section 16.14 in H&C.

2.   (a) Octahedral; (b) tetrahedral; (c) trigonal bipyramidal; (d) tetrahedral.

3.   Dissociation of the complex can give free CO that may coordinate to the iron centre in haemoglobin; see Box 4.4 in H&C.

4.   See Section 5.12 in H&C.

5.   (a) See Figure 16.5.1; (b) tetrahedral; octahedral; (c) two, *trans-* and *cis.*

---

**Problem set 17.1**

1.   (a)   (b)   (c)

  (d)

  (e)

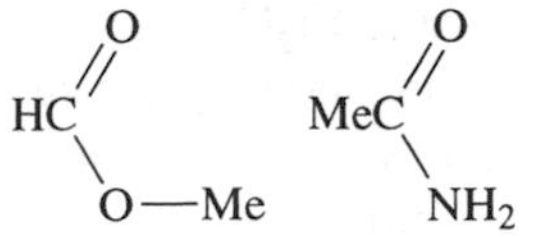

**17.1.1** Structures of methyl methanoate and acetamide.

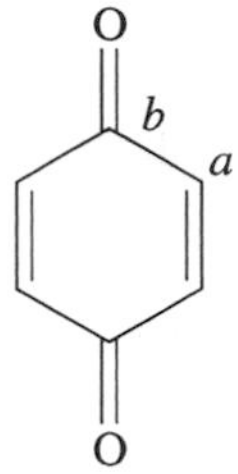

**17.1.2** The structure of cyclohexane-1,4-dione.

2. Dimer formation due to hydrogen bonding; see structure **17.24** in H&C.

3. Structures shown in Figure 17.1.1; the amide can form intermolecular hydrogen bonds resulting in association in the liquid state that raises the boiling point.

4. (a) **A** = acetamide (an amide); **B** = octyl acetate (an ester); **C** = propanoyl chloride (an acyl chloride); (b) hydrolysis occurs readily producing HCl.

5. (a) Ketone; (b) ≈ 1700 cm⁻¹ C=O stretch; absorptions 2850–3000 cm⁻¹ C–H stretches; bands below 1500 cm⁻¹ lie in the fingerprint region.

6. (a) See Figure 17.1.2. (b) Using site labels in figure: $a = \delta +136$ (rel. integral 4); $\delta = d +187$ (rel. integral 2).

7.

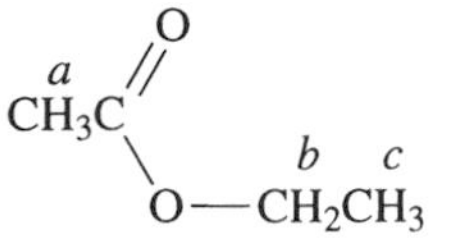

$a$ = singlet, 3H;
$b$ = quartet, 2H;
$c$ = triplet, 3H.

$a$
$CH_3C$
$b$   $c$
$O—CH_2CH_3$

8. $CH_3CH_2C(O)NH_2$.

---

**Problem set 17.2**

1. (a)       (b)

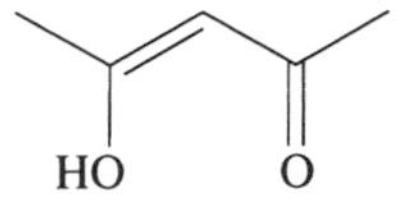

O   O        O   OH        HO   O

2. (a)

O   O⁻        O   O        ⁻O   O

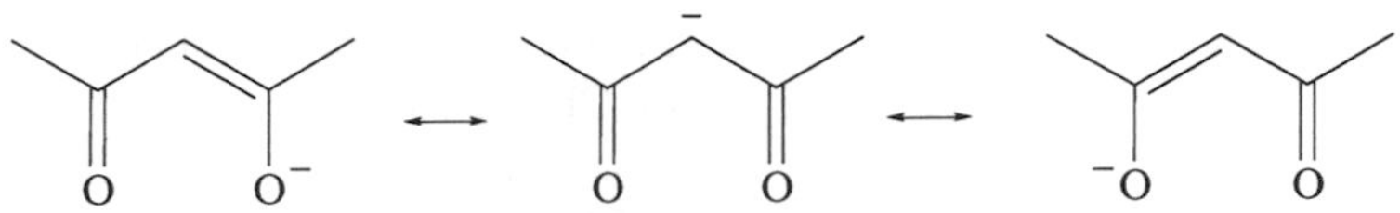

(b) 6; (c) there are $^n/_2$ resonance forms when $n$ electrons are involved.

➤ **Answer 2(c) only holds for *acyclic* delocalized systems**

3. 3

4. (a) The *position* of the equilibrium in question 1(a) is solvent dependent. Both tautomers shows signals due to Me ($\delta$ 20–30) and oxygen-attached $sp^2$ C ($\delta$ 190–200). The *keto*-form predominates in solvent **A** with a signal for the CH₂ carbon at $\delta$ 58; the *enol*-form predominates in solvent **B** with a signal for the =CH– carbon at $\delta$ 100. (b) ¹H NMR spectrum or observation of $J_{CH}$ in the ¹³C NMR spectrum.

---

**Problem set 17.3**

1. Inductive effect (see Section 14.2 in H&C) results in an increase in p$K_a$ (i.e. increase in acid strength) as the number of Cl substituents increases.

2. (a) $Cs[PhCO_2](aq)$; (b) $Na_2[O_2CCO_2](aq) + CO_2(g)$; (c) $[NH_4][CH_3CH_2CO_2](aq)$    [+ $H_2O$ in each.]

3. (b) and (c); in (b), each H is an α-hydrogen atom; in (c), the CH₂ group contains two α-hydrogen atoms.

4. (a) β-diketone; (b) $[RC(O)\bar{C}HC(O)R]$; β-diketonate; (c) formation of the coordination complex $[Fe\{RC(O)CHC(O)R\}_3]$.

5. NaOH; $CH_3C(O)CH_2C(O)OCH_2CH_3$ is a weak acid, and requires a strong base to remove the α-hydrogen atom.

6. In $RC(O)CH_2C(O)R$, α-hydrogen atom is adjacent to *two* C=O groups; the conjugate base is stabilized by delocalization of the negative charge:

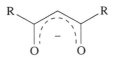

**Problem set 17.4**

1. Oxidation using (a) Corey's reagent; (b) e.g. acidified $K_2Cr_2O_7$.

2. (a) Butyl butanoate; (b) an equilibrium mixture of reactants and products is obtained; drive to the right-hand side (i.e. products) by removing ester or water by distillation, or by adding an excess of alcohol or acid.

3. (a) $Li[AlH('Bu)_3]$ at 195 K; (b) acidified $K_2Cr_2O_7$; (c) $C_2H_5OH$ (esterification); (d) $H_2O$; (e) $PCl_3$, $PCl_5$ or $SOCl_2$; (f) hot alkaline $KMnO_4$; (g) NaOH or $NaHCO_3$.

4. (a) Reaction of 1-chloropentane with Mg in anhydrous $Et_2O$ or THF gives $CH_3CH_2CH_2CH_2CH_2MgCl$; treatment with dry ice (solid $CO_2$) gives $[CH_3CH_2CH_2CH_2CH_2C(O)O]^-[MgCl]^+$; protonation gives hexanoic acid; (b) see Section 17.7 of H&C; (c) product would be 2-methylpentanoic acid; (d) reaction with NaCN to give $CH_3CH_2CH_2CH_2CH_2CN$, followed by acid or base hydrolysis.

5. (a) $PhCO_2H$ and MeOH; (b) $MeCO_2H$ and $PhCH_2CH_2OH$; (c) $PhCH_2CO_2H$ and MeOH; (d) $MeCO_2H$ and $Me_3COH$.

6. (a) Appearance of a broad band $\approx 3200-3600$ cm$^{-1}$ (O–H stretch); (b) this gives propanamide and a broad absorption around 3300 cm$^{-1}$ due to N–H stretch will appear; (c) appearance of a strong absorption $\approx 1700$ cm$^{-1}$ (C=O stretch); (d) as part (c); (e) *disappearance* of a sharp absorption $\approx 2000$ cm$^{-1}$ (C≡N stretch); *appearance* of a strong absorption near 1700 cm$^{-1}$ (C=O stretch) and a broad band near 3200–3600 cm$^{-1}$ (O–H stretch).

7. (a) $CH_3CH_2OH$: triplet and quartet with equal $J_{HH}$ values, rel. integral 3 : 2; broad signal (OH); $CH_3CO_2H$: sharp singlet and a broader signal at low field; (b) $CH_3C(O)Cl$: singlet; $CH_3C(O)NH_2$: singlet and broad signal (N–H); (c) $CMe_3CH_2C(O)Cl$: 2 singlets (rel. integral 9 : 2); $CMe_3CH_2CHO$: 3 singlets (rel. integral 9 : 2 : 1) with the lowest intensity signal at lowest field.

**Problem set 17.5**

1. Polar bond giving $C^{\delta+} O^{\delta-}$; see the beginning of Section 17.13 in H&C.

2. Haloform reaction; see Section 17.13 in H&C.

3. (a) Catalyst; in the RDS, H$^+$ protonates the carbonyl O atom, resulting in the formation of the *enol*-tautomer after H$^+$ loss; (b) $I_2$ reacts with the enol in the fast step of the reaction; (c) see Figure 7.5.1.

4. (a) Strong bases; [EtO]$^-$ and [$^i$Pr$_2$N]$^-$ ; (b) deprotonation of the α-hydrogen atom followed by:

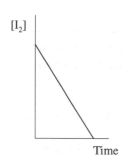

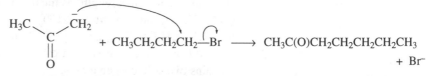

5. $MeC(O)CHEtC(O)CH_2CH_3$ and $MeC(O)CEt_2C(O)CH_2CH_3$; alkylation at the carbonyl carbon atom.

6. $H_2O$

7. See Section 17.13 in H&C.

8. (a) Primary amine of the type $RR'CHNH_2$; (b) $NH_3$ and $H_2$ / Ni catalyst; product is 1-propylamine; mechanism: see Section 17.13 in H&C.

9. See Section 17.13 in H&C.

**17.5.1** Plot of [$I_2$] against time for the acid catalysed iodination of acetone.